THE HIMALAYA

THE HIMALAYA
Aspects of Change

A Selection

Edited by J. S. Lall

DELHI
OXFORD UNIVERSITY PRESS
BOMBAY CALCUTTA MADRAS
1995

Oxford University Press, Walton Street, Oxford OX2 6DP
Oxford New York
Athens Auckland Bangkok Bombay
Calcutta Cape Town Dar es Salaam Delhi
Florence Hong Kong Istanbul Karachi
Kuala Lumpur Madras Madrid Melbourne
Mexico City Nairobi Paris Singapore
Taipei Tokyo Toronto
and associates in
Berlin Ibadan

First published 1981
Oxford India Paperbacks 1995

ISBN 0 19 563574 4

Printed in India at Rekha Printers Pvt. Ltd., New Delhi 110020
and published by Neil O'Brien, Oxford University Press
YMCA Library Building, Jai Singh Road, New Delhi 110001

PREFACE TO THE PAPERBACK EDITION

Change in the Himalaya, ever more palpable, ever more disturbing, has absorbed Himalayanists the world over. These changes are manifest in the natural heritage itself, in the often dramatic changes in the life of Himalayan people, and once again in the interrelations between the two.

An opportunity to present studies of the processes at work came when I was Director of the India International Centre in the 1970s. Studies grouped in these three categories, by well-known specialists from within and beyond the subcontinent, were brought together in a single volume. Its reception over the years has encouraged the publishers and me to hope that a selection from the original book would foster the interest that inspired the concept twenty years ago. These concerns, we are convinced, have become more relevant because of the concentration of the pace of change in recent years.

It is a concern that has mounted and widened in scope. Sir Edmund Hillary has returned to the Himalaya in a new role after his pioneering ascent of Everest with Tenzing Norgay in May 1953. Tenzing, a man of the Himalaya, was but one of its many sons and daughters achieving a new synthesis with the wonders and perils of their heritage.

It is also now a more caring one, binding together a new generation, a new body of readers, to whom this paperback edition is presented. It is our hope that they will find in it pleasure, information and cause for commitment to conserving all that the Himalaya stand for.

J. S. L.
September 1994

CONTENTS

Maps

CONTRIBUTORS

SÁLIM ALI (d. 1986), ornithologist, was President of the Bombay Natural History Society, Bombay

K. C. SAHNI was formerly Director, Biological Research, at the Forest Research Institute, Dehradun

M. K. RANJITSINGH was Senior Regional Advisor of the United Nations Environmental Programme, Bangkok, Thailand

H. M. CHAUDHURY was Director (Seismology) at the Indian Meteorological Department, New Delhi

CHRISTOPH VON FÜRER-HAIMENDORF was Professor Emeritus of Asian Anthropology, University of London

J. S. LALL was Diwan to the Chogyal of Sikkim and former Director of the India International Centre, Delhi

D. D. BHATT was Professor at the Centre for Economic Development and Administration, Tribhuvan University, Nepal

B. N. GOSWAMY was Professor and Head of the Department of Fine Arts, Panjab University, Chandigarh

K. L. KHOSLA was the Surveyor General of India

H. C. SARIN was President of the Indian Mountaineering Foundation, Delhi

GYAN SINGH was Principal of the Himalayan Mountaineering School, Darjiling, India

HELENA NORBERG-HODGE is the Director, The Ladakh Project, and recipient of the 1986 Right to Livelihood Award

B. K. ROY BURMAN was Head of the Centre for Rural Studies, Visva-Bharati Sriniketan, West Bengal, and is Chairman of the Government of India Committee on Cooperation in Tribal Areas

I. K. BARTHAKUR was formerly Director of the Economic Census (*ex-officio*), and Head of the Directorate of Economics and Statistics, Arunachal Pradesh, India

INTRODUCTION TO THE PAPERBACK EDITION

The Himalaya, whose austere majesty makes them seem eternal, are nevertheless everywhere subject to change. Nature itself has left incomplete the collision of plates which first heaved them skyward. There is thus a hidden potential of dramatic natural change—yet nature's hand has been slow, moderate and gradual, healing its own scars. Nature and early man, the few who sought refuge there or quiet for meditation, achieved a benign accommodation, harmonizing the slow processes of natural and social change without major upheavals.

But in the Himalaya, as everywhere else, man has now become the principal agent of change. At some point of time, impossible to determine precisely, discordance arose. Man, acquisitive and extractive by nature, multiplied in number and extended the range of his needs. He sought more from the mountains than they could yield without disturbing the balance of nature. The years immediately following World War I would seem to mark a turning point; till then hillmen had always been more insulated from change than people in the plains below. The wheel and motor were only just beginning to make their appearance, yet population pressure had kept mounting all the while. The effects of this were apparent to administrators in the 1930s: cultivation was pushing into the forests and marginal land; a growing population everywhere pressed upon a limited resource.

The departure of the colonial rulers ushered in a new era of development which dramatically accelerated the pace of change. Vast new projects, the wonders of Nehru's India such as the Bhakra Dam, became the pride of India's new nationhood. Engineers fought their way up the valleys building roads as far as the highest mountain passes. Patient human labour and the occasional stick of gelignite gave way to whole battalions of workers and batteries of machines. Governments were in a hurry. The mountain system could not accommodate the suddenness of these changes, which were sometimes conducted with the thoroughness of a blitzkrieg.

Conceding that the combined effects of man's struggle for survival and the penetration of the Himalaya in the cause of development are unavoidable, it is apparent that they must be reconciled with preservation of the ecosystem.

It is a problem of the greatest magnitude for it affects not just the 50 million people in the Himalaya and the 250 to 350 million in the plains skirting the mountains, but in a wider sense the future of the whole region.

It is the land and forests that have suffered most. For those who have walked in the Himalaya for days on end without meeting a soul it will come as something of a shock to be told that there is a problem of population at all, which in fact is much acute than in the plains.

The critical relation is the number of people supported by the land compared to the net area sown. Taking the two states of Himachal and Uttar Pradesh into consideration, the density of population per hectare of cultivated land in the Himalayan area is nearly four times greater than in the plains. This accounts for the growing proportion of men in the hill districts of Uttar Pradesh being compelled to seek a livelihood elsewhere. A study of Bhikiasen block disclosed that as much as 60 per cent of family income was generated in the plains. To an increasing extent human survival depends on a money order economy.

There is no way of measuring the damage to the soil caused by deforestation and the extension of cultivation resulting from increasing population. Since it is the forests which hold soil together, control run-off and conserve water, erosion loss has magnified exponentially in the short time span of the last fifty years. The cumulative effect of Himalayan erosion has been dramatically revealed by a satellite photograph taken in 1974 of the Bay of Bengal. Charged heavily with silt, Himalayan rivers, combining to form the Hooghly, are creating a new land mass, 50,000 square kilometres in extent, about 700 kilometres from its mouth. Natural erosion, a hundred times magnified by contemporary man-made erosion, is carrying soil scooped out of the Himalaya far out to sea. It has been estimated that, left to itself, nature takes 500 to 1,000 years to create one inch of top soil. Millions of hectares in the Himalaya have already been irretrievably lost. If present trends continue unabated a major tragedy surely lies ahead.

The power generated by descending rivers was dramatically highlighted by the glacial flood in the Lachung river in Sikkim in 1950. A huge chunk of solid ice collapsed into the lake just below the Sebu La, tearing apart the moraine barrier and sending almost the entire contents of the lake hurtling down the valley. From calculations made at that time this avalanche of water took about fifteen minutes to descend from about 5,500 metres to 3,000 metres at Lachung. Fortunately the village is a couple of hundred metres above the river; and, as the disaster happened at night, most villagers were in their homes, thus escaping the twelve-metre high wall of water which went roaring down the valley. A rock the size of a double-storeyed village house was deposited in the sand above the river bed. With true Buddhist understatement, the villagers called it the new guest.

This is the kind of water power which no mathematical formulae can begin to estimate. It should provoke sobering thought for planners hoping to utilize Himalayan resources.

The socio-economic consequences of change have been greatly aggravated by the policy of isolation followed by the British right up to the end of their rule. While the lower hills were exposed to developments in the plains, and the middle hills rather more selectively so, the highland communities in large areas were excluded from the range of settled administration and all that normally followed. The ferment of ideas had scarcely reached them. This applied particularly to almost the whole of present-day Arunachal Pradesh as well as large areas of Himachal, highland Uttar Pradesh, Sikkim and Ladakh. When development eventually reached them, people in the border areas were all the more vulnerable to the harsh cultural shock which hit them with a force whose consequences have not yet been fully absorbed.

Their way of life was buffeted by yet another chance development external to themselves, and this was much more abrupt and more immediately cataclysmic. When the Chinese occupied Tibet the old freedom of trans-border movement was restricted and the 1962 war finally brought trade and communications between India and Tibet to an end. The highland communities in India were ruthlessly thrown on their own resources. Some special quality of spirit bred in the high mountains enabled them not merely to survive but to adjust heroically to the traumatic break in their life styles.

Geologists attribute the structure of the Himalaya to tectonic forces, but the Indian mind can be forgiven if it sees a supernatural hand in the design of a bow tensed between the matchless beauty of two of its greatest mountains. This comparison with a bow can be extended further, for the immense mountain system consists of three distinct laminations, so to speak, consisting of parallel longitudinal ranges. They were formed by three successive thrust movements that heaved them up from the Tethyan sea. In geological time the earliest, and the highest, is the Great Himalaya, with an average altitude of about 6,000 metres. Along it are strung no fewer than thirty peaks soaring above 7,300 metres, and 5,220 glaciers. This towering range is an effective barrier against witheringly cold blasts blowing across the Tibetan plateau; moderating in turn the winters in north India and creating the monsoonal system which is the lifeblood of its agriculture. The Great Himalaya thus divide two strikingly different climatic systems from each other—the steppe and cold deserts of Euro-Asia and the monsoon-fed plains of the Indian subcontinent.

Along the edge of the Great Himalaya formed by the Central Thrust is the broader band of the Middle Himalaya, a tangled mass of valleys and ranges, with major rivers cutting through in deep gorges. The Middle Himalaya range in height between 2,500 and 4,000 metres. Between them and the plains are the Outer Himalaya, or low foot-hills known as Siwaliks, running along the entire range at an average height of 900 to 1,500 metres. The Siwaliks are broader and more distinctly formed in the west, and virtually disappear in Eastern Nepal and Sikkim only to reappear in Arunachal Pradesh.

Hidden from immediate view are two major thrust faults, along with some minor ones, which together result in frequent earthquakes of moderate to high intensity. Attempts to guard against the seismic consequences have given rise to intense controversy between hydroengineers and ecologists, so far without moderating the planning of multi-purpose projects on a grand scale, such as the Tehri dam.

A noticeable feature of the Himalayan system is the close correspondence between its geological and climatic aspects. Various consequences follow. For the traveller, the most obvious is the abrupt drop in temperature and humidity during the ascent from valley bottom to mountain top. No less apparent is the change in vegetation, from scrub and hardwood trees in the Siwaliks to pine in the middle Himalaya, merging with oak and a distinct belt of deodars in the west and silver fir in the east. These in turn give way to spruce in the west and rhododendrons and juniper in the east, skirting permanent snow. Less noticeable until the upland pastures are reached are carpets of alpine flowers, more trans-Caucasian in character in the western half of the range than in the east. Here primulas and snowdrops are the first to appear from the receding mantle of snow. Here too troops of choughs perform dazzling aerobatics, flights sweep through of the rare grandala, and, in the widening streams, white-capped redstart dart among the rocks.

Nature is much more varied in the middle valleys. Mingled calls are heard of barbets, cuckoos and colonies of babbling thrush, while the fluted whistle of the metallic blue Himalayan whistling thrush echoes in the wooded valleys. Animals have learnt to take cover from predatory man and are therefore less easily seen. Only rarely now does the snow leopard pounce on its prey of mountain goat, and blue sheep pause nervously alert on open crags.

The effect of the enormous difference between the great height of the mountain system and that of the alluvial plains, in most cases within hardly two hundred kilometres, is dramatically illustrated by the character of water flow. Beyond the Great Himalaya, the Indus, flowing from the Kailash range to the west, and the Tsangpo (Brahmaputra) flowing eastward, descend gradually from about 4,500 metres to about 3,500 metres at Leh and somewhat lower near Lhasa. But when they cut through the range, the descent through gorges thousands of metres deep becomes precipitous, generating devastating power. In Baltistan, the Indus roaring down its narrowing gorge is called 'lion river'.

Rivers originating in the southern face of the Great Himalaya and intermediate ranges, such as the Dhauladhar in Kashmir and Himachal Pradesh, descend at even greater speed till they broaden out through the Siwaliks. Two main consequences follow: an extremely high rate of erosion, especially in the more fragile eastern Himalaya, and heavy flooding during the three monsoon months of concentrated rainfall. The incidence of floods and the devastation caused thereby has increased appreciably in recent decades. This trend has been aggravated by continuous depletion of the vegetational cover for commercial and domestic use.

Although there is a clear need to mould development plans to local requirements, official agencies tend to rely on their standard models instead. Few examples are more glaring than the adoption of monoculture in the nineteenth century in a misguided zeal for commercial returns. Natural forests were replaced in valley after valley by the hardy and tenacious pine. Bio-diversity has practically disappeared from all but a few isolated places beyond the reach of ubiquitous departmental officials. The harm done to the Himalaya, and the people who depended on natural forests for a variety of needs, is incalculable, and in practical terms irreversible.

It is obvious that mega multi-purpose projects, which exercise an irresistible fascination over politicians and planners alike, only aggravate the acute problem of the survival of the Himalaya as a natural system and a place for human settlements. It is often argued, with apparent conviction, that mega and medium projects benefit the rest of the country without undermining the integrated system of Himalayan resources, and that the human problem is adequately met by re-settlement schemes. So great is their appeal for planners that run-of-the mill and small to medium projects are given scant consideration. Nor is attention focused on more efficient management of existing schemes, which are estimated to be about thirty per cent underutilized. Adding to them will only compound the problem of salination which has already reached dangerous proportions in some areas. Of course, there is nothing specially glamorous about increasing the efficiency of existing schemes. It is so much easier to spend thousands of crores now on new schemes than to worry about their relatively distant negative effects.

In the Himalaya, if any one resource holds the key then surely it is the forests. Over one hundred years ago, in 1864, the Commissioner of Meerut wrote: 'The district still abounds in fine trees, from one hundred to two hundred years old and upwards. All these fell before the axe [the railways were being built at a feverish pace], and probably the rest would have gone had the roads been any better.'

Extension of communications has not proved an unmixed blessing.

If change is both desirable and inevitable, surely there must be a methodology of change consistent with the preservation of all that had inspired the sage Nagasena? He spoke thus of the Himalaya:

> The Himalaya, the king of the mountains, five and three thousand leagues in extent at the circumference, with its ranges of eight and forty thousand peaks, the source of five hundred rivers, the dwelling place of multitudes of mighty creatures, the producer of manifold perfumes, enriched with hundreds of magical drugs, is seen to rise aloft, like a cloud, the centre (of the earth).

What poet could match the splendid measure of this ancient hymn which so grandly orchestrates the sentiments of today's mountaineers, ecologists, foresters, hydro-engineers, botanists, animal lovers and conservationists?

In the truest sense, it sets the tone of the pages that follow.

CHAPTER 1

THE HIMALAYA IN INDIAN ORNITHOLOGY

SÁLIM ALI

The Himalayan mountain range is a gigantic physical barrier which cuts off the Indian plains from the high plateaux of Tibet and Central Asia. On the one hand it bars the northward passage of the moisture-laden south-west monsoon currents, and on the other it serves to insulate the plains of northern India from the severity of the continental climate of Asia—the scorching desiccating winds of summer and the icy gales of winter. These factors are responsible for the marked differences in the fauna and flora of the northern and southern aspects of the mountains; an abrupt change in the environment strikes the observer immediately upon his crossing the main Himalayan axis. The ornithologist soon misses such familiar companions of his trek up to this point as the Whitecapped Redstart (*Chaimarrornis leucocephalus*), Whistling Thrush (*Myiophonus caeruleus*), Snow Pigeon (*Columba leuconota*) and Alpine Chough (*Pyrrhocorax graculus*), and suddenly finds in their place species not seen before. The range constitutes the boundary between two of the six zoogeographical regions of the earth, namely the Palaearctic (Europe, north Africa stretching across north and central Asia) and the Oriental Region south of this, which is subdivided into the Indian, Indochinese and Indomalayan subregions.

Mean temperature drops at an average rate of 1° C for every 270 m of ascent, the drop being steeper and more rapid above 1500 m. Thus the lofty Himalaya encompass the entire spectrum of temperature, from tropical heat near the base—in the *tarai, bhabar* and *duars,* and in some inner valleys—to arctic cold at the heights, through the gradation of subtropical, temperate and alpine, the arid areas above being under perpetual snow. This wide diversity in physiographic conditions gives rise to altitudinal belts or 'life zones', corresponding closely to the climatic zones of latitude in continental Asia, except that on the high mountains these zones are telescoped vertically instead of being horizontal. Though diffusing into one another at the seams, these life zones are often sufficiently clear-cut in their flora and fauna for the altitude to be roughly predictable without the aid of an aneroid. In fact one of the joys of trekking in the Himalaya for the observant naturalist is the changing kaleidoscope of the biota as he ascends from one altitudinal belt to another. In the alpine zone of the high Himalaya which corresponds to the arctic tundra, scrub replaces forest above the limit of tree-growth, and depending upon the degree of moisture, only certain xerophytic annuals and perennials grow along with the dominant lichens and mosses—as in the tundra. The snow-line, because of the difference in latitudes (e.g. Srinagar *c.* 34° N, Thimpu *c.* 27° N) is lower in the western Himalaya than in the eastern; in summer between *c.* 4500 and 5500 m, in winter down to 2450 or 2750 m.

The altitudinal zones of vegetation, particularly in the eastern Himalaya, are of the highest interest to the student of bird ecology inasmuch as each of them harbours a more or less characteristic avifauna of its own. Perhaps nowhere in the world would one find so much diversity in climates and vegetation telescoped into so circumscribed a space. Hooker gives a dramatic example of Sikkim: 'From the bed of the Ratong, in which grow palms with screw-pine and plantain, it is only seven miles in a direct line to the perpetual ice. . . . In other words, the descent is so rapid that in eight miles the Ratong waters every variety of vegetation from the lichen of the poles to the palm of the tropics; whilst throughout the remainder of its mountain course, it falls from 4000 to 300 ft flowing amongst tropical scenery, through a valley whose flanks rise from 5000 to 12000 ft above its bed.'

Geologically speaking, the Himalaya are of recent origin and it is unlikely that the avian endemics confined to this mountain barrier date further back than the Pleiocene, and possibly came considerably later. As suggested for the High Himalayan insects by Mani (1956), evidently the avifauna also represents, at least in part, a geographical relict fauna of the Pleistocene of central Asia. Mani remarks, furthermore, on the deep penetration of the range by insect forms of the tropics and subtropics up to relatively high altitudes—a phenomenon associated with local microclimates and humidity consequent on the penetration of the tropical evergreen forest types pointed out by Hooker. From entomological and botanical evidence it seems valid that the climatic conditions in the Himalaya during the Pleistocene glacial period were not too severe, and permitted the existence and distribution of animals and broad-leaved trees such as the horse-chestnut (*Aesculus indica*). The Himalaya thus acted as a *refugium* and many animal species found there today do, in fact, represent Palaearctic relicts.

The Oriental element in the avifauna is richly represented in the eastern Himalaya and gradually diminishes westwards until in Kashmir and farther west it ceases to be a significant constituent, its place being taken by Palaearctic forms. The largest number of genera of birds of the forested area from Arunachal Pradesh to south-east Kashmir occurs also in the hills of western China and northern Burma but not in peninsular India or the Palaearctic. The infusion of Yunnan and Szechwan avifauna is strongest in the eastern Himalaya and west to central Nepal. The Kali Gandaki river in central Nepal (not Arun-Kosi in eastern Nepal as previously thought) has been shown by the Flemings (1976) to be the dividing line between the avifaunas of the western and eastern Himalaya. In floristics too, western Himalayan flora differs from eastern Himalayan in a greater representation of conifers; while European elements are conspicuous in the former. Malayan, Chinese and Burmese elements are in the latter.

The Himalayan barrier encouraged the spread of tropical Indochinese related bird species into the Indian subregion and prevented the invasion of Eurasian-related species except largely as winter migrants.

The distributional radiation of birds in the Himalaya has evidently been from western China and Assam and not from peninsular India. Blanford's explanation of the process seems eminently valid. He says: 'When glacial snow and ice spread, the tropical fauna which may at that time have resembled more closely that of the Peninsula was forced to retreat to the base of the mountains, or perished. Assam valley and the hill ranges south of it afforded a securer refuge—being damp, sheltered and forest-clad—than the open plains of north India and the drier hills of the country south of these. During the glacial period the Palaearctic forms survived in greater numbers since a considerable percentage of the genera recorded from the Tibetan subregion of the Palaearctic are not found in the Indomalayan subregion. As glacial conditions receded the Oriental fauna re-entered the Himalaya from the east.'

Bird families endemic to the Himalaya, not found in peninsular India, are Broadbills (Eurylaemidae), Honey-guides (Indica toridae), Finfoots (Heliornithidae), and Parrotbills (Paradoxornithidae). Two monotypic genera of the Pheasant family (Phasianidae) are endemic to the Himalaya, namely *Catreus* (Chir Pheasant) and *Ophrysia* (Mountain Quail). Both are considered by Ripley (1961) to be Palaearctic relicts. A third endemic genus *Callacanthis*, a cardueline finch, is also not found elsewhere besides the Himalaya where its only species *burtoni* occurs in wet and moist temperate forest between elevations of 1800 m and 3000 m.

In addition to these, Ripley (1961) shows that some fourteen Palaearctic endemic species are confined to the Himalaya, without adjacent relatives, and give strong evidence of being relict forms. As notable among these he mentioned the Wood Snipe (*Gallinago nemoricola*), Himalayan Pied Woodpecker (*Picoides himalayensis*), Blackthroated Jay (*Garrulus laneeolatus*), Smoky Leaf Warbler (*Phylloscopus fuligiventer*), Pied Ground Thrush (*Zoothera wardi*), Crested Black Tit (*Parus melanolophus*), Beautiful Nuthatch (*Sitta formosa*), and two species of Bullfinch (*Pyrrhula erythrocephala* and *P. aurantiaca*).

Other interesting forms, though not occurring in adjoining Tibet nor in the Indian subcontinent south of the Himalaya, extend more or less along the entire range. While it would be tedious to give a bare listing, there are certain groups and individual species which merit special mention since they lend character and distinction to the avifauna of the Himalaya.

THE BIRDS

The avifauna of the Himalaya, thus, is mainly a conglomerate of Palaearctic and Indochinese elements, the former predominating in Kashmir and the western section, the latter in the eastern area. This is not the place for a catalogue of all the birds to be found in the mountains. Such a list is available in several of the more recent references cited at the end. There are, however, a number of families, or groups of birds, or individual species, which are either peculiar to the Himalaya or are so narrowly associated with the mountain range as to call for special notice. Some of these birds are of special interest not only for their aesthetic appeal but because so little scientific data are available concerning their status and ecology on account of their living at heights beyond the ceiling of the average ornithologist, and of their rarity and restricted habitats. It is to be hoped that this account will help to fill the gaps for observant trekkers and mountaineers. Two such examples among the raptors or Birds of Prey would indisputably be the Golden Eagle (*Aquila chrysaetos*) and the Lämmergeier or Bearded Vulture (*Gypaetus barbatus*). The magnificent Golden Eagle is a large and powerful hunter, dark chocolate-brown (almost black) with a tawny rufous hindcrown, nape and hindneck. It usually keeps in pairs, frequenting desolate rugged mountain country with crags and precipices, between *c*. 1850 m and summer snow-line (3000–5500 m). The species, in a number of other geographical races, or subspecies, ranges widely over the Holarctic Region, south to north Africa and the USA. The Himalayan subspecies, *daphanea,* also spreads to Turkestan and from eastern Iran to central Asia. Its prey consists of pheasants, snowcock, chukor, hares and occasionally larger mammals such as musk deer and the young of mountain sheep.

It is difficult to understand why the Lämmergeier is so commonly referred to as Golden Eagle even by otherwise fairly knowledgeable sportsmen. Except that they are both huge spectacular raptors found in similar high Himalayan biotopes and nowhere else within the Indian subcontinent, there is little justification for confusing the two. The Lämmergeier is less massive than the Golden Eagle. It has a fully feathered cream-coloured head and neck, silvery grey-and-black upper parts, and is rusty white below. In sailing flight the comparatively narrow, pointed wings and longish wedge-shaped tail are diagnostic, while the 'goatee' of black bristly feathers at chin, specially conspicuous in profile, clinches its identity. The so-called Bearded Vulture is a purely mountain-living form found in the Himalaya between *c*. 1200 and 4000 m and sometimes seen sailing majestically even above 7000 m. The same subspecies, *aureus,* inhabits the mountains of south-east Europe (Pyrenees, etc.), south Arabia and eastward to north China. Two other subspecies occur: one in the mountains of north Africa (Algeria, Morocco, etc.), the other in Ethiopia to south Africa. The Lämmergeier is best known for its habit of carrying aloft large bones, like the femur of an ox, and dropping them on rocks, often at regular 'ossuaries', from a height of 50 m or more, in order to splinter them. The bird then descends to feed on the fragments of bone and the marrow, which largely comprise its diet.

On many counts the family Phasianidae—Pheasants, Junglefowl, Partridges, Quails—collectively and popularly known as 'Game birds'—may be regarded as the most distinctive bird family of the Himalaya. The Himalayan pheasants, whose centre of distribution is primarily the Indochinese subregion of the Oriental Region, include some of the most fascinating, spectacular and gorgeously-plumaged birds in the world. They are so eagerly sought by private collectors, aviculturists and zoos everywhere, but particularly in the western countries, that ruthless exploitation for profit, chiefly by poachers and smugglers, has reduced the population of several species to dangerously low levels, threatening their extinction in the wild. Happily, most pheasants breed freely in captivity, and it is hoped that by re-introduction of captive-bred stock into seriously depleted areas it may be possible in some measure to rehabilitate the more vulnerable species. Poaching, as well as large-scale habitat destruction for the construction of roads, and other dubious forms of economic development, is the main cause for the declining number of this remarkable group of birds. The genera of Himalayan pheasants concerned are mainly *Ithaginis* (Blood Pheasant), *Tragopan* (Horned Pheasant), *Lophophorus* (Monal or Impeyan Pheasant), *Lophura* (Kaleej Pheasant), *Pucrasia* (Koklas Pheasant) and *Catreus* (Chir Pheasant). But the Eared Pheasant (*Crossoptilon*) and Peacock Pheasant (*Polyplectron*), though somewhat marginal, also merit a rightful place among the avifauna of the Himalaya.

The genus *Ithaginis*, Blood Pheasant, with the single species *cruentatus*, and numerous geographical subspecies, is distributed in the eastern Himalaya and associated mountain ranges of China, from Nepal to Kansu. In not extending westward beyond Nepal, the Blood Pheasant graphically exemplifies the Chinese influence in the eastern Himalayan avifauna. It is a bird of the highest altitudinal zone among the true Himalayan pheasants, affecting steep pine forest with rhododendron, juniper and ringal bamboo scrub near the snow-line, alternating seasonally between *c*. 3600 and 4300 m altitude. Three subspecies are recognized within Indian limits. The Blood Pheasant is a large, gaudily-coloured partridge-shaped bird, chiefly grey and apple-green, with a mop-like crest, black forehead and crimson throat. A black-bordered bright red naked patch surrounds the eye. The upper breast is splashed with crimson, like fresh blood-stains, and there are similar splashes also on the shoulders and the tail. The female—bright rufous-brown with ashy-grey crest and nape —is so startlingly different in looks that initially it was even described as a different species from the cock. Successive observers have remarked upon the 'stupidity' of members of a covey coming out in trustful inquisitiveness to a companion fluttering in its death throes, and thus being killed one by one by the ambushed hunter—a shameful commentary on the vileness of man rather than on the bird's stupidity!

The genus *Tragopan,* Tragopans or Horned Pheasants have the shape, proportions and carriage of partridges. The males have a great deal of resplendent crimson in their plumage. The sides of the head and throat are naked and brilliantly coloured in most species. In addition, they possess two long brightly-coloured fleshy horns, one above each eye, which are erected during courtship display. A brilliantly coloured and patterned fleshy apron-shaped bib-like wattle on the throat is fully expanded at the same time, heightening the bizarre effect. There are four species within Indian limits, which live at altitudes between *c.* 1400 and 4250 m, descending to slightly lower levels in severe winters. The Western Tragopan, *T. melanocephalus* (Punjab Himalaya, Himachal Pradesh, Garhwal) is probably the most threatened species at the present time and needs extreme protection. It is included in the Red Data Book of the IUCN and also in Schedule 1 (totally protected species) of the Wildlife (Protection) Act, 1972 of the Government of India.

The genus *Lophophorus* is represented in the Himalaya by two species of remarkably beautiful pheasants, namely the Impeyan or Himalayan Monal (*L. impejanus*) and Sclater's or the Mishmi Hills Monal (*L. scalteri*). They are stoutly built, dumpy birds shaped like the Snowcock (*Tetraogallus* spp.), but distinguished by the highly refulgent bronze-green, purple and blue plumage of the males. The face is more or less naked and coloured bright blue. The two species are readily distinguished from each other by the shape of the crest. In the Impeyan the crest is composed of feathers with naked shafts and a spatulate end; in Sclater's it is of ordinary feathers which are short and curly. The females of both are rather similar: brown, mottled and streaked with paler and darker colours, with a pure white throat and a bare pale blue patch round the eyes. They have a short crest of ordinary feathers. These pheasants live in high oak, rhododendron and deodar forest with open glades and sheep pastures, at between 2500 and 5000 m altitude. While the Impeyan is confined more or less entirely to the Indo-Pakistan Himalaya, Sclater's inhabits the section in Arunachal Pradesh and the contiguous parts of southeast Tibet, eastward to north-east Burma and Yunnan.

The commonest and most abundant genus of the true Himalayan pheasants is *Lophura* to which belongs the Kaleej (*L. leucomelanura*) together with a number of extralimital Indomalayan species. They are all distinguished especially by the long, pointed, arching tails of the cocks which are laterally compressed and 'roof-like', as in the domestic and junglefowl (*Gallus*)—their closest allies. Kaleej may further be identified by their coloration which is chiefly a glistening metallic-black in the male, and reddish-brown in the female. The latter lacks the characteristic sickle-shaped tail of the cock. In both sexes there is a bushy backwardly directed crest and naked bright scarlet face. Within the Indian Himalaya five (or six) races of the Kaleej are recognized from north-west Pakistan to Arunachal Pradesh, and in the associated Assam hills. Popularly known as *Kālā Mūrghā,* Kaleej share their native habitat with Red Junglefowl, particularly at the lower altitudes. They are found in all types of forest—sal, oak, rhododendron, spruce, etc.—from *c.* 400 up to 3600 m. They live amidst heavy scrub and undergrowth, and like junglefowl, are partial to the neighbourhood of water and terraced cultivation.

The genus *Pucrasia,* Koklas Pheasant, contains a single species *macrolopha,* found more or less along the entire length of the Himalayan system, from Afghanistan to eastern China. The species is enigmatically absent east of Nepal, but after a considerable break in the distribution it reappears farther east in Yunnan and the mountains of China and Mongolia. Four geographical races, or subspecies, are recognized in the stretch between Afghanistan and Nepal, based on minor differences of coloration and plumage pattern. Both sexes have the face fully feathered and with a well-developed occipital crest. The cock has two long tufts of metallic-black plumes on either side of the crest, springing from the ear-coverts, which are erected in nuptial display. The overall coloration is largely grey, black and chestnut, and the sexes are dimorphic. The Koklas are seen in wooded ravines on steep hillsides in oak and conifer forest with heavy scrub and ringal bamboo undergrowth, within an altitudinal range of 1500 to 4000 m.

The genus *Catreus* contains a single species, the Chir Pheasant (*C. wallichii*). It is endemic to the Himalaya and confined to a restricted range from north-west Pakistan (Hazara) to west-central Nepal (Pokhara between *c.* 1400 and 3500 m elevation). It lives on steep rugged hillsides in oak forest covered with tall grass and scrub and dissected by wooded ravines. It is an extreme skulker, taking to its legs when hunted, and reluctant to fly unless flushed by a dog. The closely-barred pale rusty upper parts, long backward-projecting crest of hair-like plumes, bright crimson naked patch round the eye, and above all the long pointed tail broadly barred with black and grey, provide a good identification key for the cock. The hen is rather similar but has a shorter tail.

To make this list of pheasants comprehensive, it seems pertinent to add two other genera, even though their inclusion among the true Himalayan avifauna may be somewhat equivocal.

The genus *Crossoptilon,* Eared Pheasant, is predominantly Chinese. Of the three species known, only one (*C. crossoptilon*) enters Indian limits on the extreme north-east fringe of Arunachal from its main range in adjoining south-east Tibet, north of the Himalayan axis. The Eared Pheasant is a large, heavy and showy bird chiefly blue-grey in colour with a laterally compressed metallic blue-black tail, the long arching distintegrated central rectrices recalling ostrich plumes. The crown is velvety black, and the sides of the head naked and deep scarlet. The ear-coverts are prolonged to protrude behind like horns (or ears), hence the bird's popular name. The sexes are alike, but the male is larger. Our subspecies *C. c. harmani,* Elwes's Eared Pheasant, lives at altitudes of between 3000 and 5000 m, on grassy hill-slopes at the edge of rhododendron and juniper scrub. Where unmolested, as in the

neighbourhood of Buddhist monasteries, the birds become astonishingly tame
and even come to be fed by the monks.

Finally, we have that elegant and quietly beautiful genus *Polyplectron,* the
Peacock Pheasants, of smallish birds reminiscent in their shape and bearing
of Spurfowl. Several species are found in the Indochinese countries and
Malaysia but only a single one, *bicalcaratum,* with two subspecies, in the
north-eastern part of the subcontinent—Sikkim, Bhutan, Arunachal, etc. It
is a bird of grey, grey-brown and buff plumage, studded with brilliant metallic
violet, green-blue eye-spots or ocelli on the mantle and fan-shaped tail. The
male has a short frowzled upstanding crest. The hen is smaller and duller
coloured, with shorter crest and tail. The Peacock Pheasant is a low-elevation
bird, normally not found above *c.* 1200 m. It lives in the dense tropical foot-
hill jungle of the Himalayan complex and is an inveterate skulker, keeping to
heavy undergrowth and almost impossible to flush even with a dog.

One of the most elusive, intriguing and little-known members of the
Pheasant family, and of the entire Himalayan avifauna, is the Mountain
Quail, *Ophrysia superciliosa.* This bird has defied all attempts to rediscover it
since the last specimen was collected in 1876. It was originally described for
science in 1846 from a living pair, supposedly from 'India' that had found its
way to a private menagerie in England. Not until its discovery in the wild
around Mussoorie, many years later, was the exact provenance of the species
determined and the type-locality fixed. In all there are only ten museum
specimens of the Mountain Quail extant, five in the British Museum and the
rest in America. The only locality in which the species is known is the
Kumaon Division of the western Himalaya, all the specimens having come
from elevations between 1650 and 2100 m—Jharipani (a little below Mussoorie), Benog and Bhadraj peaks behind Mussoorie, and Sher-ka-danda
peak near Naini Tal. It was conjectured at the time that the Mountain Quail
may be a migrant, but since the bird has not been recorded from any neighbouring country or anywhere else in the world, its origin and status continue
to remain a mystery. It is hard to account for its extreme rarity unless the
species is already extinct. The taxonomical position of *Ophrysia* is somewhat equivocal. The structure of its wing is reminiscent of the Spurfowl
(*Galloperdix*); the short stout bill, long lax plumage and the stiff bristle-like
feathers on the forehead are features that recall the Blood Pheasant (*Ithaginis*); the bill and forehead bristles are also very much as in the Bush Quails
(*Perdicula*). In size the bird is between a Grey Partridge and a Grey Quail,
but the tail is proportionately longer than in either. The sexes are dissimilar.
The male is dark slaty-brown overall, with distinctive black-and-white face
markings and a broad white eyebrow; the female is largely cinnamon-brown
with a pinkish grey face and a less prominent white eyebrow. The bill and
feet in both sexes are red. The birds are reported by early collectors to live in
coveys of five or six in tall grass from which they could be flushed only when
almost trampled on. They pitched into the grass again after a short heavy

flight—and are therefore in behaviour also very similar to the bush quail.

Among the many other groups and species of birds peculiar to the Himalaya or which have special interest and significance in the context of Himalayan ornithology, the Orange-rumped Honeyguide (*Indicator xanthonotus*) is one that merits immediate attention. It is a rare species, and little is known of its ecology. The family Indicatoridae (Order *Piciformes*) closely related to the Barbets (Capitonidae), is African in domicile, being represented in that continent by three (or four) genera and thirteen species. One of these genera —*Indicator*, with two species—has a curiously disjunct transcontinental distribution in south-eastern Asia: *I. xanthonotus* occurs in the Himalaya between *c.* 1500 and 3500 m from the Afghan frontier right across to Arunachal and Manipur, the other species *L. archipelagicus* being confined to Malaysia. In the Himalaya it affects mixed broad-leafed and conifer forest in the neighbourhood of bee-nesting cliffs and rock scarps. No definitive data are available on the ecology of the Himalayan Honeyguide though by interpolation of what is known about its African cousins its general life pattern can be constructed. The name 'honeyguide' arises from the fact that by means of certain well-understood calls and actions, such as flitting from tree to tree in front of a honey-questing human or other mammal, the bird guides the seeker to a comb and shares the remnants—honey, larvae and wax—after the honey has been taken. An extraordinary anatomical adaptation of the bird is its ability to digest wax, which few other animal systems can assimilate. Thus a considerable proportion of its diet apparently consists of beeswax. In Africa, honeyguides of this genus are known to be brood-parasitic on barbets, laying their eggs in the nest-holes of the latter and leaving the dupes to foster the hatchlings. Recent observations by two American ornithologists in Nepal, though not yet fully confirmed, suggest that the Himalayan Honeyguide has a novel type of breeding biology, described by them as 'resource-based non-harem polygyny'. The male holds a bee's comb, or group of combs, as his territory throughout the year and mates with all receptive females that visit it to feed during the breeding season. These may total up to twenty or more! Males without a territory apparently seldom have a chance of mating (Cronin, Jr., and Sherman, 1977).

Ward's Trogon (*Harpactes wardi*) is found in the eastern Himalaya, from central Bhutan eastward, between 1500 m and 3000 m. This pigeon-sized bird is similar to the commoner Redheaded Trogon but is much larger (overall length 40 cm). It is brilliantly coloured; the belly is crimson-pink in the male and primrose yellow in the female. The tail is graduated and its feathers squarely truncated at the tip. The central retrices are black and the lateral ones pink in the male and yellow in the female. The bird is seen either singly or in separated pairs in subtropical forest amidst bamboo and evergreen undergrowth. It feeds on large insects and berries.

Redbilled Leiothrix or 'Pekin Robin' (*Leiothrix lutea*) is a sprightly bright-

coloured sparrow-sized bird, greyish olive with a bright yellow throat and
breast, a pale eye-ring and a scarlet bill. The wings are black with yellow-
and-crimson edges. In the female, the crimson of the wings is replaced by
yellow. It breeds between 1500 m and 2400 m and descends lower in winter.
It is seen in pairs or parties of four to six or even twenty, affecting evergreen
biotope: secondary growth, overgrown clearings and tea plantations. Its
food consists of insects, berries and seeds. Its harsh hissing notes and loud
cheerful bulbul-like warbling are diagnostic.

Redbilled Blue Magpie (*Cissa erythrorhyncha*)—A showy Himalayan bird
found between *c*. 1000 and 2000 m is the Redbilled Blue Magpie. It can be
seen from *c*. 77° E in Himachal Pradesh (Kangra dist.) eastward through
Garhwal and Kumaon to *c*. 87° E in the eastern Himalaya. Its reported
occurrence in western Sikkim needs verification. It is very similar to and
confusable with the Yellowbilled Blue Magpie, but it usually occupies a lower
altitudinal zone and is oftener met with at Himalayan hill stations. In Mussoorie, UP, it is known as *Nilkanth* and in Simla, *Digdal*. It is a spectacular,
purplish-blue bird with a long graduated tail. The central feathers are elongat-
ed into gracefully arching streamers. The head, neck and breast are velvety-
black with a large white patch on the nape. The underparts below the breast
are greyish white. Both sexes are alike.

THE HIMALAYA AND BIRD MIGRATION

The importance of the Himalaya in the context of Indian bird migration is
only just beginning to be properly appreciated. Of the 2100 odd species and
subspecies of birds that comprise the avifauna of the subcontinent together
with Sri Lanka, nearly three hundred are winter visitors from the Palaearctic
Region north of the Himalayan barrier (Eurasia to north-east Siberia and
central Asia). This excludes the purely pelagic forms such as petrels, shear-
waters and boobies which sometimes get accidentally blown in on the sea-
board by monsoon gales. Many of these migrants are actually Himalayan
species that breed up to altitudes of 5500 m or more in summer, e.g. Red-
starts (*Phoenicurus*), and move down to lower levels in the foot-hills or to the
north Indian plains in winter, some continuing southward to spread through
the peninsula and even spill over into Sri Lanka. Among our most regular
and abundant winter visitors, the 'classic' long-distance extra-limital migrants,
are the hordes of ducks and geese (Anatidae), cranes (Gruidae) and the vast
assortment of wading or shore-birds (Charadriidae) which become conspicuous enough on every *jheel* and wet-land to obtrude their presence on the
least observant. But the smaller birds, in spite of their overwhelming numbers
and far greater variety, are not usually noticed unless specially looked for.
The passerine families such as Muscicapidae (flycatchers, warblers, thrushes,
robins), Motacillidae (wagtails and pipits), Fringillidae (finches), Hirundinidae
(swallows) and some others are amongst our most abundant winter visitors.

Visual observations recorded over a series of years by British sportsmen–naturalists, mostly army officers posted in the strategic north-west frontier areas in the early part of the nineteenth century, had suggested the broad pattern of the seasonal migratory movements, particularly of sporting birds like ducks and cranes, and established the Indus Valley as the putative principal flyway of Siberian and central Asian birds into peninsular India. Less substantial but similar evidence from the north-eastern end of the Himalaya suggested that the main route on that side was along the valley of the Tsangpo or Brahmaputra river, and its affluents. The two migrational streams entered from either end of the Himalayan mountain chain in a sort of pincer movement, weakening in their advance as more species dropped out, before reaching the tip of the peninsula or trickling into Sri Lanka, which is the virtual terminus for our land birds.

Until recently, it was commonly assumed that the Himalayan range presented an insuperable barrier to migrating birds from northern lands, and that the only way they were enabled to cross it was by following river valleys. It used to be argued that the freezing cold and rarefied atmosphere at the formidable heights which the birds would need to negotiate would overpower them. Thus, western students of bird migration had believed that migration flight normally took place under about 500 m, rarely as high as 1500 and only exceptionally around 3000 m while crossing lofty mountain ranges. Sophisticated devices and vastly improved technology between the two world wars, and especially the invention of radar and refining of relevant scientific techniques have, however, radically altered the older notions. It is now well established that a considerable amount of bird migration everywhere does occur beyond the range of human vision, and that even small birds, of starling size, when migrating, may fly at previously unsuspected heights of 6000 to 7000 m, braving the intense cold, even when there is no obvious compulsion for them to do so. The growing popularity of Himalayan mountaineering since the First World War, and the virtual rash of climbing expeditions since the Second World War, have adduced convincing evidence that a very considerable—maybe even overwhelming—proportion of the migration to and from the Indian subcontinent occurs directly over the main Himalayan axis, thereby appreciably shortening the birds' journeys. Autumn migration in particular, by direct overflight would appear to involve little difficulty since the birds would merely need to coast down, as it were, from the elevated tableland of central Asia to the Indian plains, with a maximum saving of time and energy. For the upward return journey in spring, perhaps the river-valley route could have more validity in certain sections. That this is a hazardous undertaking nevertheless, exposing birds to the unpredictable vagaries of the notorious Himalayan weather such as sudden storms and blizzards, sometimes resulting in mass annihilation, is evident from the reports of explorers and mountaineers from time to time. The famous explorer Sven Hedin observed large numbers of migrating ducks in autumn at great heights at the source of the Indus river in Tibet. The first Mount Everest Expedition (Young-husband's) came across several birds on southward migration in September at *c.* 5200 m, among them Temminck's Stint (*Calidris temminckii*), 'Painted Snipe' (?),Pintail Snipe (*Gallinago stenura*), House Martin (*Delichon urbica*), and several pipits (*Anthus* spp.). More than once, migrating waders were heard passing overhead at night at this altitude, among which the Curlew (*Numenius arquata*) was unmistakable. Others have recorded, visually during the day and by ear at night, migrating geese, cranes and waders passing over their high-altitude camps. Dr Biswamoy Biswas of the Zoological Survey of India noted a stray hoopoe (*Upupa epops*) on Pumori Glacier (5790 m) in May, which had no business to be there unless on migration. Blacktailed Godwits (*Limosa limosa*) and Pintail Duck (*Anas acuta*) have been observed on Khumbu Glacier (*c.* 4875 m) at the foot of Mt Everest. Colonel R. Meinertzhagen came across various species of duck in Ladakh on passage to India over the highest parts of the Himalaya. Eric Shipton on his 1937 expedition to the Karakorams found a large number of dead, frozen ducks and possibly a crane strewn over the face of the Crevasse Glacier between *c.* 4500 and 4900 m and in the upper basins of most of the big glaciers he visited. C.H. Donald, a competent Dharamsala-based west-Himalayan naturalist in the twenties, frequently recorded large numbers of migrating geese and cranes, in one case both Common and Siberian, flying regularly in spring and autumn over the passes from Ladakh, over Chamba and the Lahul ranges, at between *c.* 4500 and 6000 m; and more recently in Nepal, R.L. Fleming observed Barheaded Geese (*Anser indicus*) flying north in spring over the eastern spur of Dhaulagiri at *c.* 7625 m.

In May 1960 Brigadier Gyan Singh, the leader of the first Indian Mt Everest Expedition, found three Eastern Steppe Eagles (*Aquila nipalensis nipalensis*) lying dead on the South Col—'One of the most difficult areas to cross'—at a height of nearly 7925 m, which had obviously perished while on migration. Since the Everest climber Tenzing also mentions in his autobiography that he saw a dead eagle on the South Col in autumn 1952 when with a Swiss expedition, it would appear that the South Col is on the regular flyway of these eagles between India and central Asia. That birds can exist and fly at immense heights, seemingly with little physical discomfort from the intense cold and low atmospheric pressure, is suggested by one of the Everest expeditions encountering crows and mountain finches around their high-level camp at 7000 m, and Himalayan Griffon Vulture and Lämmergeier between 6000 m and 7000 m, while Alpine Choughs (*Pyrrhocorax graculus*) even followed the climbers up to 8200 m, where the atmosphere is reduced to only one-third of its supporting power. Sir Edmund Hillary has reported a chough following him at 8500 m—presumably one of the several individuals that scavenged daily around their camp at 7900 m. At Dehra Dun geese have been observed through a telescope, flying northward in spring across the face of the moon at a calculated height of 8830 m.

Altitudinal Migration

Apart from the host of long-distance trans-Himalayan migrants there is a large number of species that breed at high altitudes and descends to lower levels or into the northern plains in winter, some indeed extending well into peninsular India and even Sri Lanka. Of those that stay permanently in the mountains and merely move from a higher to a lower elevation, a typical example is the Grandala (*Grandala coelicolor*). The male is a gorgeous purple-blue mayna-like bird, while the female is chiefly brown, streaked with white. The grandala lives in large flocks and breeds between 4300 and 5500 m in summer, seldom descending lower than 3000 m in winter. In particularly severe weather however, it is sometimes forced down as low as 2200 m, the lowest limit recorded. A better-known though not too common example is the brilliantly crimson-winged Wall Creeper (*Tichodroma muraria*), an ashy-grey sparrow-sized bird with slender longish bill and butterfly-like flight, which lives on high cliff-faces around altitudes ranging from 4300 to 5500 m, and breeds within rock fissures. In winter the Wall Creeper descends to the base of the mountains and is often seen in quarries and rock cuttings in the adjoining plains. This bird has even been observed as far south as New Delhi, clambering on the walls of the Central Government Secretariat, no doubt having mistaken that august edifice for Mount Olympus!

There are certain Himalayan endemics that evidently fly non-stop on their annual migrations to the hills of the southern–western Ghats or Sahyadri complex. Prominent among this group are the Woodcock (*Scolopax rusticola*), Pied Ground Thrush (*Zoothera wardii*), Blue Chat (*Erithacus brunneus*), Brownbreasted Flycatcher (*Muscicapa muttui*), Bluethroated Flycatcher (*Muscicapa rubeculoides*) and several other flycatchers. All of them seem to be species of low tolerance and find the requisite ecological conditions only in the damp well-wooded southern hills which, as mentioned earlier, already support several relict forms of Himalayan flora and fauna. These species, whose migration routes evidently lie along the eastern and western ghats, are fairly common in the south Indian hills during winter but are very seldom met with on passage in the intervening country. It is thus evident that they must perform their migration to and fro in a single hop, involving a minimum distance of maybe 1500 to 2000 km each way. Among the more typical of such single-hop migrants is reckoned the Woodcock (*Scolopax rusticola*). This is a coveted game bird—a dumpy overgrown snipe—much sought after by sportsmen in its Himalayan homeland as well as in its southern winter-quarters—the hills of the Western Ghats complex, traditionally the Nilgiris.

HIMALAYAN RELICTS IN SOUTH INDIA

An intriguing aspect of Himalayan flora and fauna is the disjunct occurrence of so many typical genera and closely-related species, as endemics in the hills of south India. Among mammals two outstanding examples are the Nilgiri Tahr (*Hemitragus hylocrius*) and the Nilgiri Pine Marten (*Martes gwatkinsi*), but birds provide many more examples. The Laughing Thrush genus *Garrulax* (family Muscicapidae, subfamily Timaliinae) whose centre of distribution is the Indochinese subregion of the Oriental, is strongly represented in the eastern Himalaya, and found throughout the mountain range westward to Kashmir and Afghanistan in something like 27 species. After a total absence of over 2000 km in continental and peninsular India, the genus reappears in the south with two endemic species, *cachinnans* and *jerdoni,* the former confined to the Nilgiris, the latter in three well-differentiated subspecies, to the Palnis and Kerala hills. There is, additionally, a third member of the genus found in Kerala, namely the Wynaad Laughing Thrush which is an obvious subspecies of the east Himalayan *Garrulax delesserti.* A point of special interest in the case of these far-flung laughing thrushes is that their presence in the southern hills is closely associated with the occurrence of the plant genus *Rubus* (raspberry, blackberry, etc.) which itself is a common Himalayan taxon with an identical disjunct distribution in the hills of the southern peninsula. The close ecological association between *Garrulax* and *Rubus* in the Himalaya is faithfully reproduced in their common southern refugium. Thickets of *Rubus* first make their appearance in the south Indian hills from an altitude of about 1000 m upward, an elevation at which, significantly, *Garrulax* also appears on the scene. Other examples of this sort of symbiosis are not uncommon. The disjunct occurrence of so many Himalayan animals and plants in the south Indian hills presents some intriguing problems of logistics. The late Dr Sunder Lal Hora, a distinguished Indian ichthyologist, found several genera of specialized torrential fresh-water fishes in Kerala which were identical with those in the eastern Himalaya and the Malay peninsula. Since the geographical distribution of fish is inescapably dependent on a direct connection at one time or another (river capture) of the streams and rivers they inhabit, and they can spread in no other way, Dr Hora after careful research propounded a rational explanation in his 'Satpura Hypothesis for the distribution of Malayan fauna and flora to peninsular India'. While this related primarily to Malayan–east Himalayan highly-adapted torrential fresh-water fishes, it seems to answer satisfactorily for other organisms as well. The hypothesis has received varying support from several different disciplines and investigators, but still remains equivocal and calls for further intensive in-depth studies. It seems relevant here to highlight some of its salient points. The Satpura Hypothesis postulates that the Satpura–Vindhya trend of mountains stretching across India was at one time more elevated and more humid than now. It was continuous with the Assam hills in the east and with the northern end of the Western Ghats (in Gujarat) in the west, and thus served as a causeway for the spread of specialized Himalayan flora and fauna into the south Indian hills and Sri Lanka. The link of the Assam Himalaya with the Satpura–Vindhya trend lay across the Garo–Rajmahal gap on the Chota Nagpur plateau, which was postulated to have been more raised and humid than now, providing a continuous stretch of ecologically-suitable hilly country as a bridge for the dispersal of plants and animals. Geotectonic processes such as erosion or subsidence of the connecting highland and consequent changes in the physiography may be responsible for cutting off the continuity in the distribution of the ecologically-adapted organisms now met with as 'marooned' relict populations of Himalayan animals and plants in the southern hills. Geological evidence on the whole is not in favour of this easy explanation, and alternative possibilities need to be investigated. Nevertheless, the Satpura Hypothesis offers an attractive interpretation which, on biogeographical evidence alone, seems unexceptionable.

Among the more prominent Himalayan relicts in the avifauna of the Western Ghats complex in southern India, besides the Laughing Thrushes (*Garrulax*), are the Great Pied Hornbill (*Buceros bicornis*), Frogmouths (*Batrachostomus*), Fairy Bluebird (*Irena*), Lizard Hawks or Bazas (*Aviceda*), and Rufous bellied Hawk-Eagle (*Hieraaetus kienerii*). Several others also show the same widely disjunct distribution and are palpably relict Himalayan forms.

PIONEERS OF HIMALAYAN ORNITHOLOGY

Most of our early knowledge of Himalayan avifauna comes from Brian Houghton Hodgson (1800–94), the undisputed 'Father' of Himalayan ornithology. Hodgson first went to Nepal at the age of twenty and was British Resident in that then difficult country from 1833 to 1843. During a period of about twenty years, in spite of the constraints on free movement for foreigners, he managed to collect over 20,000 skins of birds and mammals which he later presented to the British Museum. He published numerous scientific papers which form the hard core of our knowledge of Himalayan birds. After retirement from Nepal he lived and continued his natural history work from Darjiling between 1845 and 1858, when he finally left India. Hodgson (1833) was the first to write on bird migration in India, of ducks, etc., as observed by him at Kathmandu, and also the first to draw attention to the altitudinal distribution of species in the Himalaya ('Physical Geography of the Himalaya', 1840).

Of the many other naturalists who have contributed to Himalayan ornithology before and since Hodgson, perhaps the one deserving special mention is John Gould. Gould was the taxidermist in charge of the Zoological Society's museum in London, and later superintendent of its ornithological collection. He was a gifted bird painter and, in collaboration with his artist wife, published in instalments, commencing in 1831–2—partly on the basis of Hodgson's material—a splendid folio of coloured lithographs entitled *A Century of Birds of the Himalayan Mountains*. The scientific text for this was written by Nicholas Vigors, the Secretary of the then newly-founded Zoological

Society of London, who was responsible also for the description of many of Hodgson's newly-discovered species from Nepal.

BIBLIOGRAPHY

ALI, SALIM. 1949. *Indian Hill Birds*. Bombay. Oxford University Press.

——. 1949. 'The Satpura Trend as an Ornithogeographical Highway'. *Proc. Nat. Inst. Sci. India.* Nov.- Dec.

——. 1962. *The Birds of Sikkim*. Madras. Oxford University Press.

——. 1977. *A Field Guide to the Birds of the Eastern Himalaya*. New Delhi. Oxford University Press.

CHAMPION, HARRY G. and S.K. SETH. 1968. *A Revised Survey of the Forest Types of India.* Nasik. Government of India Press.

CRONIN, EDWARD (Jr) and PAUL SHERMAN. 1976. *Living Bird.* Cornell University.

FLEMING, R.L., Jr. 1971. 'Avian Zoogeography of Nepal'. *The Himalayan Review* 4: 23–33.

FLEMING, R.L., Sr. and Jr. and LAIN SINGH BANGDEL. 1976. *Birds of Nepal.*

HODGSON, B.H. 1833. *Asiatic Researches.*

HOOKER, J.D. 1905. *Himalayan Journals.* London. Ward, Lock & Co.

HORA, S.L. 1937. 'Distribution of Himalayan Fishes and its Bearing on Certain Palaeogeographical Problems'. *Rec. Ind. Mus.* 39: 251–9.

——. 1949. 'Symposium on Satpura Hypothesis of the Distribution of Malayan Fauna and Flora to Peninsular India'. *Proc. Nat. Inst. Sci. India* 15(8): 307–422.

——. 'Hora's Satpura Hypothesis: An Aspect of Indian Biogeography'. *Current Science* 19: 364–70.

SECRETARY OF STATE FOR INDIA. 1908. *Imperial Gazetteer of India.* Oxford. Clarendon Press.

MANI, M.S. 1956. 'Insect Fauna of the High Himalaya'. *Nature* 177: 124.

MEDLICOTT, H.B. and W.T. BLANFORD. 1879. *A Manual of Geology of India.* 2 vols. LXX: 374. Calcutta.

MEINERTZHAGEN, R. 1928. 'Some Biological Problems connected with the Himalaya'. *Ibis.* 480–533.

RIPLEY, S.D. 1961. *A Synopsis of the Birds of India and Pakistan.* Bombay. Bombay Natural History Society.

STEPHENSON, J.S. 'The Geographical Distribution of Indian Earthworms'. *Proc. Asiat. Soc. Bengal* (N.S.) 12: cxvii.

WADIA, D.N. 1939. *The Geology of India.* London. Macmillan.

CHAPTER 2

BOTANICAL PANORAMA OF THE EASTERN HIMALAYA

K. C. SAHNI

The eastern Himalaya, comprising North Bengal, Sikkim, Bhutan and Arunachal Pradesh is literally a botanist's paradise. Phytogeographically it forms the meeting ground of the Indo-Malaysian (S.E. Asian) and Sino-Japanese (E. Asian) floras. The eastern end of the Himalaya has been a gateway for the migration of plants from E. and S.E. Asia. Even a cursory examination of Merrill's *Plant Life of the Pacific World* (1945) shows the similarity between the flora of the foot-hills of the eastern Himalaya and those of the Pacific region. Birbal Sahni (1939) at the Sixth Pacific Science Congress, on the basis of palaeogeographical, geological and palaeobotanical data suggested 'The eastward opening of the Himalayan Geosyncline into the Pacific Ocean'.

The mingling of species in this region has possibly favoured natural hybridization with subsequent enrichment of the variability of flora. The lower reaches of Arunachal, Bhutan and Sikkim comprising the evergreen rain-forests teem with plant life and have a bewildering number of species and genera. In fact, this region including the old Assam State is perhaps the richest botanical region of the world and has been a favourite hunting ground for plant explorers like Griffith (1847), Hooker (1849), Burkill (1924), and in later years Bor (1938), Biswas (1941), Kingdon-Ward (1960), Rao and Panigrahi (1961), Hiroshi Hara (1963), Sahni (1969), and Deb and Dutta (1971). In recent years Japanese botanists have explored the eastern Himalaya, including Bhutan (Hara, 1966, 1971). Officers of the Botanical Survey of India have been active in Bhutan (Deb, 1968; Anonymous, 1973) and in Arunachal and Assam (A.S. Rao, 1974). Botanists from the Forest Research Institute have contributed on Sikkim (Sahni, 1960) and Arunachal (Sahni, 1962, 1969).

In the following account, besides the geography, climate and flora, the impact of man and technology and the methodology aimed at in preserving and enriching the environment of this unique area is briefly discussed.

GEOGRAPHY AND CLIMATE

Sir Joseph Hooker, the great naturalist of the mid-nineteenth century was the first to make known the beauty and natural history of Darjiling and Sikkim after his significant traverses in the eastern Himalaya. Facing the Ganga delta, the Tista valley reproduces on a vast scale. The upper basin is 80 km wide and the river cuts through the Darjiling ridge (2130–2430 m) in a narrow gorge to spill on to the plains in a vast fan. With a rainfall of 3040–4570 mm, erosion is intense. Darjiling, at 2248 m, has a climate typical of the outer hills. Mist-enshrouded for half the year, on clear days it boasts a magnificent view of Kanchenjunga and a less spectacular view of Everest which is far distant. The range of relief in Sikkim is 330–8579 m or over 8 km. At dawn from Singhik there is a breath-taking panorama of Kanchenjunga (8579 m). To the west the state is shut in by the Singalila ridge, to the east by the Donkia Range (4570–5540 m). Chho Lhãmo lake is the source of the Tista and lies in the shadow of Donkia Mountains. It is a calm blue sheet of water hemmed in by rounded spurs from Mt Khangchengyao (6934 m)—rocks and glaciers dip abruptly towards its head. The mountains bordering Chho Lhãmo plateau are a rusty orange colour which is caused by peroxide of iron. The remarkable features of this landscape are its enormous elevation (Chho Lhãmo, 5250 m, Donkia La, 5540 m), and undulating hills rising to 5560 m, bare of snow. Extraordinary dryness and consequent evaporation increased by strong winds account for the height of the snow-line. Bordering Tibet there are vast gently undulating, scantily grassed, sandy plains like the cold deserts of Ladakh. Here, because of a strong sun, sterile soil and radiated heat a piece of snow does not melt but disappears by evaporation. In west Sikkim, Mt Siniolchu, 6890 m, is reputed to be the most beautiful mountain in the world. In the outer hills rain is almost continuous from June to September. Above 1830 m, snow falls almost every year. The upper reaches of the river Tista are known as Lachen and Lachung, which have their confluence at Chungthang, and form the Tista which is the mightiest drainer of the eastern Himalaya.

Bhutan is a country of wild mountains and forests and rolling yak pastures. In its north-west corner it is dominated by the superb peak of Chomo Lhari, 'Divine Queen of the Mountains' (7314 m), guarding the Tang La col on the Brahmaputra/Tsang-po watershed. The main rivers are Ama Chu and Manas which both drain into the Brahmaputra.

In Arunachal Pradesh the Himalaya constitute about 70,000 sq. km of the total area of about 80,000 sq. km. The Arunachal Himalaya extend in a north-easterly direction through the north-bank districts of Kameng, Subansiri and Siang, and then make an orographic swerve south into the Lohit, and eastern part of Tirap districts, where the sub-Himalayan belt becomes insignificant. Lohit and Tirap districts are located east of the Brahmaputra. The lower Himalayan ranges (up to *c.* 3500 m) merge into the Great Himalaya through the striking Se La Range. This sweeps north-eastwards from the Tibet-Bhutan border and divides the Kameng, Subansiri, Siang and Lohit districts from Tibet. The well-known peaks in the Great Himalaya in Arunachal are Gorichen (6538 m), Kangto (7090 m), Chomo, and the very beautiful Nyegi-Kangstan, (snow country of delight). The eastern-most, Namcha Barwa (7756 m), in Tibet, is visible from the northern boundary of Siang and is the highest peak east of Kanchenjunga.

The five Arunachal districts thus comprise an enormous stretch of territory and exhibit a very wide range of climate from the sweltering plains of Assam and the foot-hills to the arctic climate of the eternal snows. The eastern Himalayan ranges in Arunachal, rising steeply from the Assam plains, receive the full force of the summer monsoon current from June to September and there is very heavy rainfall in the lower ranges. This abundant rainfall is due to the horseshoe-shape of these mountains formed by the bending of the Himalaya in the Arunachal–Assam corner, which catches the bulk of the monsoon-bearing clouds rising from the Bay of Bengal.

The mountains of Arunachal are consequently more evenly humid than any known part of the Himalaya, and the outer ranges and foot-hills have a climate which may be termed 'tree producing'.

The annual rainfall of the plains is about 2280 mm but the rainfall is much heavier and attains 5000 mm at the base of the foot-hills. Up to 6350 mm has been recorded on Piri La which is probably the rainiest area north of the Brahmaputra. In the Tenga valley to the north of this mountain, however, only 1520 mm is deposited, while beyond Bomdi-La at Dirang, there is even less rainfall. The winter is severe at Ziro. Snowfall usually occurs above an altitude of 1824 m throughout the state (Burkill, 1924; Bor, 1938; Sahni, 1969).

The areas of heavy rainfall are on the windward side of the Himalaya. Rainfall, temperature, humidity, and snowfall play a very important role in determining the character of the vegetation.

VEGETATION

The vegetation of this area is unique—it is characterized by one of the richest flora in the world, and abounds in spectacular tree ferns, orchids, primula and blue poppies, and the tallest trees in India. It is the habitat of botanical rarities; such as *Sapria himalayana* with flowers 35 cm across and buds as large as a grapefruit, a parasitic plant facing extinction as it has only been sighted twice since its first collection by Griffith in 1836, a century later by Bor, and more recently by Deb; and *Entada,* an enormous climber with giant pods over one metre long (with disjunct distribution; also found in the western Ghats where it is reported to be 1.5 km in length— perhaps the largest climber in the world).

The percentage of endemic plants (and animals) is very high in the

Himalaya. The lower reaches are significant because they are regarded as a sanctuary of ancient flora which were spared the devastation of the ice ages. The *Magnolia pterocarpa* here is perhaps the most ancient species of living Angiosperms (Hutchinson, 1969). Takhtajan (1969) calls this area the 'cradle of flowering plants' and considers several species, like the vessel-less dicotyledon, *Tetracentron*, the *Magnolia* spp., *Holboellia*, etc., to be primitive. Endemic and primitive plants are dealt with at length in subsequent pages. The eastern Himalaya is far more evenly humid than the western Himalaya. High humidity is conducive to tree growth and consequently the timber line or the upper limit of tree vegetation in this sector goes up to 4570 m compared to 3600 m in the western Himalaya. It has a climate which is typically tree-producing. The foot-hills are characterized by tower-ing buttressed trees like *Tetrameles nudiflora, Terminalia myriocarpa, Bombax ceiba*, etc. In Sikkim a spruce of up to 67 m has been recorded. Owing to the heavy rainfall and the force of rain drops, many trees and herbs here assume a pendulous habit, e.g. *Picea spinulosa, Primula sikkimmensis,* etc. Also, most of the trees, shrubs, climbers and herbs in these forests have predominantly white or cream flowers, obviously an adaptation to attract night-flying moths which are common here. Thus the Nepal Trumpet Climber (*Beaumontia grandiflora*), *Elaeocarpus, Michelia doltsopa, Magnolia camp-bellii, M. globosa, Rhododendron formosum, R. nuttallii, R. wightii, R. dalhou-siae, R. lindleyi, Cassiope fastigiata, Rosa sericea, Viburnum* spp., etc., all have white flowers.

The altitudinal pattern covers a very wide spectrum, from rain forests to the arctic type of vegetation: from 200 to 6000 m, classified into five main types: (1) Tropical evergreen forests; (2) subtropical; (3) temperate; (4) subalpine; and (5) alpine to arctic. The classification followed is mostly after Champion and Seth (1968).

Tropical evergreen forests

Tropical evergreen forests extend from the foot-hills up to 800 m and are met with from Sikkim to Arunachal; they consist of a remarkably large number of species mostly confined to the families *Dilleniaceae, Annonaceae, Guttiferae, Myrtaceae, Myristicaceae, Palmaceae*, etc.

In the south bank districts of Lohit and Tirap, the forest is a broad-leaved evergeen rain forest of the Indo-Malaysian type with an assemblage of big trees. The tropical rain forest is a remarkable and unique type of forest. According to Richards (1952), 'with its enormous number of species [it] has acted in the past as a potential reservoir of genetical diversity and potential variability. It has been a centre of evolutionary activity from which the rest of the world's flora has originated through mutational alterations which successfully survive under the high and constant temperature of the rain forests. Evidence of many kinds indicate that our temperate floras have directly or indirectly a tropical origin of which many temperate species still bear evident traces either in their phenology or structure.' The World Wildlife Fund has rightly emphasised the need to preserve rain forests in different parts of the world.

The forests generally present a many-tiered appearance of which the top 'storey' is constituted by 'Hollong' *Dipterocarpus macrocarpus* (on the south bank of the Brahmaputra in Lohit and Tirap), *Artocarpus chaplasha, Tetrameles nudiflora*—the last-named characterized by enormous plank buttresses, a favourite nesting-tree of hornbills. These trees, not all of which are evergeen, tower above the rest of the canopy. This storey is followed by large trees of *Terminalia myriocarpa*, most striking when in fruit with a coppery red tinge; *Altingia excelsa, Chukrasia velutina, Lagerstroemia speciosa, Bischofia javanica* (bark favoured by tigers for cleaning their claws), *Bombax ceiba* (or semal), *Dysoxylum procerum, Sloanea assamica* (with black spiny fruit), etc. The middle storey is formed of gregarious species like the white-flowered *Mesua ferrea* (iron wood), *Phoebe goalparensis* and *Duabanga grandiflora*, all valuable species. This storey normally determines the economic value of the forest. The lowest storey comprises small trees, viz. *Syzygium formosum, Dillenia indica, Talauma hodgsonii* and canes, mostly *Calamus* spp., occurring in swamps and forming impenetrable thickets. Tree ferns like *Cyathaea* spp. and the large fern *Angiopteris evecta* are also seen associated with the screw pine *Pandanus furcatus*. The wild banana *Musa balbisiana* is a conspicuous feature and occurs gregariously towards the upper limit in evergreen forests at 1000 m. The most conspicuous epiphytic elements are the orchids, ferns and fern allies; amongst orchids *Dendrobium* spp. predominate, with *Cymbidium* coming next. A strikingly handsome 'bamboo orchid' (*Arundina graminifolia*) 2 m in height, is conspicuous along with *Galeola falconeri*, with deep yellow flowers, which stands out majestically amidst wild plantain in Kameng District.

Subtropical grasslands

Grasslands as a biotic climax appear at *c.* 1050 m where both subtropical evergreen forest and grassland vegetation are found to coexist. The grasslands owe their origin to the practice of shifting cultivation. The usual grasses to be seen are *Arundinella bengalensis, Saccharum spontaneum, Setaria palmifolia*, etc. *Themeda caudata* with *Cnicus* sp. forms a dominant grass at *c.* 1700 m. The principal components of the grassland are the bamboo, viz. *Chimonobambusa callosa, Cephalostachyum latifolium, Dendrocalamus hookeri*, and *D. sikkimensis*. The valuable broom grass *Thysanolaena maxima* occurs in commercial quantities on hillsides at 900 m.

Subtropical forests (900–1800 m)

Subtropical forests (Submontane of Champion and Seth, 1968) are dominated by a *Ficus–Castanopsis–Callicarpa* association in the lower reaches and by *Schima–Castanopsis–Engelhardtia–Saurauia* association in the higher reaches. *Araliaceae* generally occur in both these associations. *Alnus nepalensis* grows extensively along river beds. On the drier aspects *Rhododendron arboreum* and *Lyonia-Pieris* with *Saurauia* predominate, whereas the deep river valleys are lined with *Albizia* and *Morus. Populus ciliata* occurs in the inner valleys in Arunachal in the Tenga valley while *P. gamblei* is rare in Kalimpong, Bhutan and in the Subansiri District in Arunachal. It is interesting that species like *P. ciliata* and *Pinus wallichiana* descend to much lower altitudes in the eastern Himalaya than in the western, both descending in the Tenga valley to less than 1500 m. Their limited descent in the western Himalaya is due to the intercalation of dry months there which raises the temperature at lower altitudes. In general, the Himalayan species descend lower in the eastern Himalaya than in the western. *Pinus roxburghii* (Chir) a wide-ranging species of the western Himalaya occurs rarely in Sikkim, more frequently in Bhutan, and reappears in Kameng District in the inner valleys near Munna (Dirang) where it is common at 1600 m. Its occurrence in the inner valleys of Kameng is interesting because this pine generally occurs in the outer Himalaya where the full force of the monsoon is felt. Rare trees are *Magnolia pterocarpa* along the river Khru in Subansiri and two conifers, *Cephalotaxus griffithii* (at 1500 m) and *Podocarpus neriifolius* (at 1000 m), both in Kameng District. Outstanding orchids are *Cymbidium giganteum, Dendrobium chrysanthum,* and *Paphiopedilum fairieanum,* the 'lost orchid' seen in Bhutan and Kameng

Temperate forests (1800–3500 m) (E. Himalayan Wet Temperate of Champion and Seth).

These are mostly temperate rain forests. The dominant families comprising the temperate forests are the *Magnoliaceae, Ericaceae, Cupuliferae* and the *Coniferae.* The valuable building timber *Michelia doltsopa* occurs at 1900–2300 m, followed by the highly ornamental *Magnolia campbellii* with snow-white flowers, leafless when in flower at 2400–2750 m. Their common associates are: *Rhododendron arboreum, Symplocos spicata, Lyonia ovalifolia, Rhododendron falconeri, R. thomsonii, R. griffithianum, Quercus lamellosa* and the rare epiphytes *Rhododendron edgeworthii* and *R. dalhousiae.* Since the conifer genera follow specific ranges of altitude it is convenient in the present account of temperate forests to enumerate the non-coniferous species as associates of the valuable conifer genera occurring in the temperate belt. *Pinus wallichiana* (Blue pine) occurs in the Tenga valley, further north in Dirang and in the forests below Tawang and also in Apatanang in Subansiri District where it is cultivated. Further east it occurs in the inner hills of Lohit up to Walong. In the Tenga valley the climate is dry, rainfall *c.* 1520 mm, and the soil is derived from quartzites and limestone. This conifer flourishes between 1520 m to 2750 m. Here the associated trees are, *Quercus griffithii*, *Populus ciliata*, *Lyonia ovalifolia*, *Corylopsis himalayana*, *Rhododendron arboreum*, *Acer oblongum*, *Exbucklandia populnea*, *Alnus nepalensis*, etc.

Associated shrubs are *Zanthoxylum armatum*, *Rubus ellipticus*, *Debregeasia longifolia*, *Caryopteris wallichiana*, etc. The common herbs are *Rumex hastatus*, *Urtica parviflora*, etc. In Apatanang (1800–2100 m) and Tawang, the rainfall is much higher and the soil is derived from Himalayan gneiss. *Pinus wallichiana* in Apatanang attains immense proportions. Girths of 4.2 m and heights of 45.6 m are seen in some specimens. Associated trees are *Acer oblongum*, Carmine cherry (*Prunus cerasoides*), *Turpinia nepalensis*, *Picrasma quassioides*. The undergrowth is dense and consists of *Phlogacanthus guttatus*, *Lasianthus biermanni*, *Brachytoma wallichii*. *Cupressus torulosa* grows on the limestone rocks of the Tenga valley and attains huge dimensions particularly near Buddhist shrines. Its main associates are *Quercus griffithii*, *Pinus wallichiana*, *Rhododendron arboreum* and *Pyrus pashia*. The forest is very open and the shrubby vegetation is confined to species like *Elaeagnus umbellata* characterized by foliage with silvery scales and edible fruits, and *Berberis asiatica*. The common herbs are *Gerbera piloselloides*, *Viola patrinii*, *Verbascum thapsus*, etc.

Tsuga dumosa: a large tree resembling the fir, characterized by small female cones about 2.5 cm long. It is common from Sikkim to Arunachal at heights of 2400 m to 3000 m. Trees of immense dimensions occur on the Se La mountain. Associated trees are *Lithocarpus pachychylla*, *L. elegans* (*Quercus spicata*), *Prunus nepalensis*, *Rhododendron falconeri*, *R. barbatum*, *Magnolia campbellii*, *Taxus baccata*, *Illicium griffithii*, *Osmanthus sauvis*, *Symplocos* sp. and the bamboo *Arundinaria racemosa* which comes up wherever the forest opens out. Split culms of this are used for roofing and its young shoots are edible. Shrubs are *Gaultheria fragrantissima*, whose leaves are fragrant when crushed, *Skimmia arborescens*, etc.

Abies delavayi: this fir grows on the northern slopes of Piri in Kameng District between 2700–3350 m. Its real home is in Szechuan, China, and its presence in Arunachal is of interest. Associated trees are *Rhododendron falconeri*, *R. barbatum*, *R. Hodgsonii*, *R. fulgens*, *R. maddenii*, *Lyonia ovalifolia*, etc. The hill bamboo *Arundinaria aristata* is very common and its culms are eaten by elephants.

Abies densa: this fir grows over wide areas from the hills of north Bengal to Sikkim through Bhutan to Arunachal at 3200–4330 m. In the upper reaches, there is an undergrowth of the tall *Rhododendron hodgsonii*.

Larix griffithiana: this larch occurs from Sikkim to Arunachal on glacial moraines from 2740 to 3350 m. It is the only deciduous conifer of India.

Picea brachytyla: a spruce hitherto not recorded from India has been identified as *P. brachytyla* (Sahni, 1962). This was collected from Mago in Kameng at 3500 m. In China, it occurs over an extensive area in Hupeh and W. Szechuan from 1500 to 2280 m, and in W. Yunnan it covers a wide range of altitudes from 2100 to 3700 m. In Kameng District it usually occurs in pure patches and is sometimes seen mixed with *Pinus wallichiana, Abies densa* as in Mago, *Tsuga dumosa* and *Taxus baccata* as in Thingpo. This find from Arunachal is an interesting new record for India and is evidence of the long-recognized affinity between the flora of the eastern Himalaya with that of W. China (Sahni, 1969).

Picea spinulosa: a tall evergeen tree, one of the tallest of the spruces, attaining a height of over 60 m. Found from Sikkim to Arunachal at 2600–3300 m. Branches are more pendulous than in *P. smithiana* of the western Himalaya.

Subalpine (3500–4500 m)

The most dominant tree here is *Abies densa* which occurs from Sikkim through Bhutan to Arunachal. At this height it is mostly a fir–rhododendron association. The other associates like *Betula utilis* and *Juniperus wallichiana* are comparatively scarce. On the upper slopes *Rhododendron hodgsonii* with papery bark, pinkish flowers and very large leaves is common, e.g. Se La (Arunachal) and Yumthang (Sikkim). *Rhododendron hypenanthum* with creamy-white flowers, *R. campanulatum,* a gregarious shrub, *R. thomsonii* with blood-red, bell-shaped flowers, *R. camphylocarpum* with creamy-white flowers, *R. virgatum, R. cinnabarinum* with cinnamon-red to brick-red flowers, *Cassiopa fastigiata, Berberis* sp., *Salix sikkimensis* are the usual shrubs. The ground flora is mostly of ornamental herbs like *Caltha palustris, Meconopsis simplicifolia, Primula denticulata, P. capitata, Anemone obtusiloba, Fritillaria cirrhosa,* and *Arisaema griffithii,* a striking cobra-like plant.

Alpine vegetation (4500–5500 m)

The vegetation of the subalpine and alpine gradually merge with the disappearance of tree growth. Typical areas are Donkia La, Chho Lhāmą, Lhonak, Jongri, Nathu La (Sikkim), Se La, Bum La and Pan̄gchen (Arunachal). Some of the striking plants in this belt are: *Rhododendron nivale,* the smallest of the rhododendrons, only 5 cm high; *Saussurea gossypiphora,* a striking woolly herb; *Thylacospermum rupifragrum* at Donkia La where it forms large, hard hemispheric cushions 30 cm across, the growth of centuries; *Arenaria musciformis* which forms cushion-like growths near the Chho Lhāmo Lake; and in the adjacent semi-arid region *Ephedra saxatilis* var. *sikkimensis* and the handsome prickly blue poppy *Meconopsis horridula* and the yak fodder *Urtica hyperborea* are common. Edelweiss (*Leontopodium himalayanum*) the national flower of Austria, is common in the alpine region. The alpine region of Nathu La which is more humid, is characterized by gregarious patches of the ornamental *Primula calderiana* with deep pinkish-red flowers and the giant rhubarb *Rheum nobile* upto one metre high. Its dried leaves are used as a substitute for tobacco and its existence is therefore threatened.

Stony deserts are seen at over 4800 m. They are usually a litter of scree and rocks encrusted with lichen with patches of *Sedum* and *Androsace* growing among the rocks. The stony desert is a more familiar feature in Tibet and is formed by the weathering of rocks caused by a wide daily range of temperature.

Maximum altitude for vegetation

Data collected indicate that there is no altitudinal limit for flowering plants except perpetual snow. One of the Mt Everest expeditions found plants of *Leontopodium* and *Arenaria* at 6096 m or even higher. F. Smythe actually found one on the rock wall of Kamet (western Himalaya) at over 6400 m. He threw it down to Holdsworth who was at the other end of his rope, but failed to catch it, and in Holdsworth's words 'the adventurous crucifer, as it probably was, was lost to Science'. *Ermania himalayensis* sets a world altitude record for flowering plants. It was collected by Gurdial Singh on Kamet (western Himalaya) at 6400 m, and is preserved in the Forest Research Institute, Herbarium, at Dehra Dun.

Affinities of the flora

Floristic studies reveal that there is a rich Chinese and Malaysian element in the flora of the eastern Himalaya with a mingling of western Himalayan and European flora. The eastern Himalaya is thus a meeting ground of the Sino-Malaysian and western Himalayan flora. In fact, the Himalaya has served primarily as a 'route' for immigration and colonization from the east and north-west, and secondarily of endemic development (Stearn, 1960). There are some areas, where topography and climatic conditions facilitate the influx of extraneous elements. Such conditions exist in Aborland which according to Burkill (1924) 'is where the two earth systems meet, a veritable node in phytogeography. If we could make the world colder from tomorrow, the plants northward of Aborland may wander in; if we could make it hotter, the Malaysian vegetation from the south might enter; if by either change we could lead a new group of plants into W. Himalaya it might advance eastwards until Aborland is attained.' The Chinese element in the flora has been studied after reference to the work by Shun Ching Lee (1935), and the Malaysian by reference to the work by Ridley (1922–5) and Van Steenis (1948–54, 1955–8).

E. Himalayan species in China are *Abies delavayi, Acer campbellii, A. oblongum, Alnus nepalensis, Bischofia javanica, Cornus macrophylla, Dillenia indica, Exbucklandia populnea, Magnolia campbellii, Picea brachytyla, Populus ciliata, Prunus cerasoides, Pyrus pashia, Quercus lamellosa, Q.*

griffithii, Terminalia myriocarpa, etc.

E. Himalayan species in Malaysia are *Altingia excelsa, Artocarpus lakoo-cha, Callicarpa arborea, Dillenia indica, Duabanga grandiflora, Elaeocarpus rugosus, Syzygium formosum, Exbucklandia populnea, Gaultheria fragrantissima, Mesua ferrea, Styrax serrulatum, Symplocos spicata,* etc. (Sahni, 1969).

The evolution of the Himalayan flora is a subject of great interest. It is highly probable that endemic species which are rich in the Himalaya arose *pari passu* with the uplift of the Himalaya and are products of the growth of the mountain system. The extensive Nunatak system in the Himalaya and the simultaneous adaptation to the changing condition may also account for the survival of the endemic flora of the alpine regions (Gupta, 1972).

The Chinese mountains, being much older, have influenced the Himalayan flora considerably and many plants from these mountains have spread westwards to the younger Himalaya. During the tertiary period a common flora must have covered the whole of West Asia including the Himalaya, China and Japan (Hara, 1966).

During subsequent epochs, when great changes took place in the topography and climate of the region, separation of the floras must have taken place. The climatic fluctuations in the North expressed themselves in the form of intermittent glacial and warm periods (interglacial). During such warmer intermissions, many plants and animals must have thrived and many of them must have vanished during each glaciation. Large-scale migration and exchange of floristic elements must have taken place with each advance and recession of the ice sheets. This would explain the presence, in the present-day high-altitude flora of the Himalaya, of diverse elements derived from various directions. These successive changes in climate and topography during the Pleistocene period not only brought in many new elements to the Himalayan flora from the north-west and east, but also disturbed the existing elements, driving them out from the old habitats. Fossil evidence indicates that the Himalayan Larch was once distributed in the north-west Himalaya, while at present it occurs from east Nepal to Arunachal and Upper Burma (Raizada and Sahni, 1960; Rau, 1974).

In the Himalaya, glaciation however did not affect the foot-hills and consequently, the vegetation of the lower belt. These lower reaches are therefore regarded as a sanctuary of ancient families of flora and fauna. Migration of flora, survival of relicts, evolution of new species by an inter-mixing of different floras and acclimatization of species from the lower altitudes, must all have had a role in determining the present-day composition and distribution of the Himalayan high-altitude flora.

NATURAL CHANGE

Erosion, landslides, floods, earthquakes and avalanches bring about changes in vegetation. All these natural forces are active in the eastern Himalaya, more particularly floods. Soils freshly exposed by natural hazards like land-slides and earthquakes support a flora which is usually retrograde. At 1000–2000 m, *Alnus nepalensis* forests colonize such areas. *Trema* sp. (on lower belts), *Populus ciliate, Fraxinus* sp., etc. are also associated. New soils get colonized by stands of Blue Pine. In temperate and alpine zones there are frequent snow slides on regular courses, on which forests cannot develop. Trees uprooted by avalanches and snow slides are commonly found on steep slopes, and strong winds follow avalanches damaging the trees in the valleys below. In the lower reaches where trees get uprooted and the canopy opens out, weeds like *Eupatorium odoratum* and *Mikania micrantha* crop up extensively.

Floods, landslides and avalanches lead to serious erosion problems. Plants commonly used in the lower valleys for checking erosion are the *kudzu* vine, broom grass, *Vitex* sp., agave, *Cestrum aurantiacum,* etc.; in the higher reaches, *Ephedra* and prostrate species of *Cotoneaster* are used.

IMPACT OF MAN AND TECHNOLOGY

The impact of man and technology has not been entirely beneficial in the Himalaya, though it has undoubtedly raised living standards and brought prosperity to the area. It is however essential that a balanced development takes place to ensure that the environment does not deteriorate any further and the spectacular flora and fauna particularly of Sikkim and Arunachal are not depleted. We must therefore preserve sizable areas abounding in rhododendrons, orchids, tree ferns, butterflies, birds and other wildlife.

Shifting cultivation or rotating agriculture, called *jhuming* in north-eastern India, is still prevalent today particularly in Arunachal and Sikkim (see chapter 24). In Arunachal it is a major problem, affecting an area of 10,75,968 ha (Jha, 1977).

With shifting cultivation soil fertility diminishes rapidly, and with the protective vegetal cover lost, there is massive soil erosion. Shifting cultivators move from place to place destroying forests, and as a consequence ecological regression takes place: there is a drastic change from high forest to low vegetal cover and plants like *Callicarpa arborea,* and aggressive weeds like *Mikania* invade on a massive scale, leading to permanent loss in productivity.

Potential for development

The forests along the foot-hills contain valuable plywood and matchwood timbers, e.g. *Tetrameles nudiflora, Terminalia myriocarpa.* Lohit District in Arunachal has one of the biggest plywood units in India in the private sector with a large investment from the People's Fund. This fund is an experiment in the development of village community forests where the estate is managed by the Forest Department as a regular Forest Reserve but the revenue from these Reserves is shared between the Forest Department and the community at a 25:27 ratio. A forest-based co-operative industry has been set up in Tirap to evolve a pattern of wood-based industries in which every tribal household will participate directly. This venture gives employment to tribals, raises their earnings, and encourages them to preserve their forests.

The miscellaneous forest, which has no economic value, is being replaced with pure as well as mixed plantations of important timber species both indigenous and exotic, since industrially-oriented plantations of quick-growing species have become a necessity for the establishment of a paper pulp industry in the Kameng District. This has naturally upset the ecological balance. The aggressive weed *Mikania micrantha* known as the 'Mile-a-minute' climber has ruined large-scale plantations, and only limited success was achieved eradicating it by spraying weedicides in low dilutions at the stage of maximum vegetative growth. The Arunachal Forest Department has supplied timber worth Rs 35 lakhs annually to Defence and Railways.

Unless development is properly regulated, the forests with their spectacular fauna and flora will be destroyed totally. There will be no hoolocks left in the foot-hill forests to greet you with their howls at light of day, no hornbills which nest in the giant *Tetrameles,* no magnificent giant butterflies; spectacular orchids may disappear as the blue *Vanda* of Meghalaya has disappeared because of the vandalism of nursery men; striking masses of rhododendrons may not exist any longer and some, like the snow-white *Rhododendron dalhousiae* rare in Sikkim and Darjiling but still common in Arunachal, may become very scarce or even extinct when Kameng District is opened up in a big way.

Rare and endangered flora

Palaeobotanical research convincingly shows that during the geological period different taxa originated, flourished and eventually became extinct. The factors which led to their extinction are still operative today. Generally they become confined to ever smaller geographical areas before they die out. An outstanding example of the successful preservation of endemic flora is the survival of the redwoods of California, which are unique trees with a life span of 3000 years, and which attain gigantic girths and heights (Sahni, 1973). They are preserved in Redwood National Park in California. For India, which has a high degree of endemism in its flora, second only to the well-known example of Australia, the need for such protection is very real: if they become extinct in India they are lost for ever. The Himalaya has the maximum degree of endemism in the subcontinent. Chatterjee (1939) has estimated that there are 3165 endemics for the entire Himalaya and it is certain that the eastern Himalaya has stronger endemism than the western because of a richer flora. The restricted distribution is due to the isolation caused by the lofty mountain ranges and the dry Tibetan plateau to the north, and the warm alluvial plains of the Brahmaputra in the south, which act as barriers to plant migration.

According to Hutchinson the *Magnolia pterocarpa* of this area is perhaps the most ancient species of living angiospermous flowering plants. Not only its floral structure, but its geographical distribution also seems to support this view. The headquarters of the *Magnoliaceae* is the Himalayan–Burma–Yunnan region. *M. pterocarpa* is seen growing along the river Khru in Subansiri District.

Takhtajan has attempted to localize the cradle of flowering plants in north-eastern India, particularly the eastern Himalaya, Assam and Burma. Some of the more familiar plants which he considers primitive are: *Magnolia griffithii* (Assam and Burma); *M. pealiana, M. gustavii* (Assam); *Manglietia* (Assam, eastern Himalaya, S. China); *Tetracentron sinense* (eastern Himalaya, U. Burma, S.W. China); *Holboellia* (Assam, eastern Himalaya, China, etc.); *Exbucklandia* (eastern Himalaya, Assam to Sumatra); *Altingia* (eastern Himalaya, Assam, Japan and China to Java); *Myrica esculenta, Alnus, Betula,* etc. *Tetracentron sinense* var. *himalayense* is a primitive vessel-less (character of Gymnosperms) dicotyledonous tree of great botanical interest recently collected by the Japanese Expedition in east Nepal and earlier by the Botanical Survey of India from Bomdila. B. Sahni (1933) described a fossil angiosperm wood devoid of vessels from the Rajmahal Hills, Bihar, and considered it both geographically and structurally nearer to *Tetracentron*.

Plants in special need of protection are:

Sapria himalayna: a near ally of the world-famous *Rafflesia arnoldii* of Malaya (which has the largest flower known, 50–90 cm across, weighing 9 kg [Lawrence, 1960]) was discovered by Griffith in the Mishmi Hills in 1836. Bor (1953) found this most fascinating botanical curiosity a century later in the Balipara Tract. This fantastic stemless and leafless parasite is found on the roots of a giant vine, *Vitis elongata,* in the *Phoebe–Beilsch-miedia–Engelhardtia* association. The flowers measure 35 cm across and the rosy-pink buds are the size of a grapefruit. The colour of the reflexed lobes of the flower is a deep crimson with the upper surface covered with yellowish papillae. It flowers towards the end of the rains. It is threatened with extinction because it has only been sighted twice since its first discovery and was recently seen by Deb. Distribution: Mishmi Hills (Lohit dist.), Aka Hills (Kameng dist.), Assam and Manipur.

Entada pursaetha: (see Vegetation).

Toricellia tiliifolia: a tree from Kumaon and Nepal whose rarity can be judged by the fact that there is no specimen of it in the Herbarium of the Northern Circle of the Botanical Survey at Dehra Dun. The Forest Research Institute Herbarium at Dehra Dun has only five specimens. Recently collected from Kameng District.

Rhododendron edgeworthii: this very ornamental epiphyte with cream flowers was collected by the author in Kameng District and has only been collected three times since Lister found it in 1875 in the Dafla Hills.

R. dalhousiae; epiphyte on *Magnolia campbellii:* a lovely ornamental shrub with large snow-white fragrant flowers uncommon now in Sikkim and Darjeeling but fairly common in Kameng District. An associate of Blue Pine–*Magnolia campbellii,* it is usually seen below Tawang and above Jung below Se La.

R. nivale: smallest of the rhododendrons, barely 5 cm tall, in alpine pastures at 5200 m.

Meconopsis betonicifolia: a striking blue poppy of the eastern Himalaya with distribution extending to south-east Tibet, Burma and Yunnan. Introduced into cultivation at Kew Gardens. Rare in the wild.

Syndiclis paradoxa: a small lauraceous tree collected only once in Bhutan. It is likely to occur in Arunachal as a rare tree.

Phyllostachys bambusoides: a Sino-Japanese bamboo of rare occurrence in the wild in Arunachal.

Coptis teeta: a curious stemless herb in temperate forests at 2000 m in the Mishmi hills, supposed to cure all sorts of diseases from eye-sores and tooth ache to a sluggish brain or liver. Rare.

Conifers

Picea brachytyla: a handsome Chinese spruce of restricted distribution in Arunachal.

Abies delavayi: a Chinese fir reported by Bor from Peri La (see Vegetation).

Cephalotaxus griffithii: a rare conifer localized in Kameng District in subtropical forests at 1000 m.

Podocarpus neriifolius: the *Oleander Podocarpus* distributed from Sikkim eastward in the evergreen forest of Assam, Arunachal and the Andamans. Of rare occurrence in Arunachal.

Orchids

In India 1300 species of orchids are known to occur. Some of their outstanding habitats are in Sikkim and Arunachal. Sikkim in particular with its marked altitudinal zonation and its range of forest types, tropical—subtropical—temperate—subalpine—alpine, is one of the richest orchid habitats in the world, with 450 species. Arunachal which has not yet been fully explored is expected to have an equally rich or even richer orchid flora because of its vastness and wide-ranging topography. Some of the important, rare and spectacular orchids are:

Vanda coerulea: a highly-prized blue orchid of the Khasi and Jaintea Hills recently recorded from Tirap District, which needs complete protection. It has disappeared from the wild due to the depradations of nursery men. Its large delicately-blue flowers bloom from October to December.

Cymbidium grandiflorum: a giant orchid with a long decurved inflorescence, 120 cm long. Flowers 7–10 cm across and 10–20 cm in an inflorescence. Apple-green sepals and petals with an ochraceous yellow lip and column speckled with purple. Distribution: E. Nepal and Bhutan at 1500–2500 m. In bloom for three months.

Paphiopedilum fairieanum: popularly called the 'lost orchid'. (Saunders, the famous orchid firm of England, offered a prize of £1000 for its first collection from the wild.) It is very ornamental with a shoe-shaped lip and wide striped sepal. Recently collected from the Saleri valley in Kameng. In Sikkim two sanctuaries have been demarcated for its preservation. Distribution: Sikkim, Bhutan and Arunachal.

Diplomeris hirsuta: highly ornamental with white flowers, growing amongst rocks and boulders. It needs protection from the disturbance caused by the present intensive road-building activity. In order to save this orchid the Sikkim Orchid Society has appealed to orchid lovers to buy only seeds and not plants. Distribution: Nepal, Darjiling, Sikkim and Bhutan.

Galeola falconeri: the tallest orchid of the Himalaya, up to 5 m in height. It stands out majestically amidst wild banana in the Kameng forests and has deep yellow flowers.

Anoectochilus sikkimensis 'Jewel Orchid': olive-green and white sepals. Dark red leaves with a velvety sheen veined with golden yellow. Distribution: Sikkim and Arunachal.

Ferns

Cyathea gigantea: the giant tree fern.

Alsophila latebrosa: a lofty tree fern.

Osmunda regalis: the royal fern.

All these require protection because of the clearance of forests in their habitat for planting industrial woods. Pith from the trunk of the first two is used in making an intoxicating beverage. *Osmunda* is of importance as the major source of fibre used in the culture of orchids.

Alpines

Primula sikkimensis; Meconopsis spp.; *Rheum;* cushion-forming *Thylacos-*

permum and *Arenaria;* and the woolly *Saussurea gossypiphora* are all a striking feature of the alpine landscape and require protection (see Vegetation).

CONSERVATION

In India through Forest Departments, the Forest Research Institute and
the National Committee on Environmental Planning and Coordination, it is
proposed that 2 to 3 per cent of the total forest area in various regions of
India be declared as nature reserves to cover important ecological types and
natural habitats of our wild life. The National Forest Policy of 1952 has laid
down that 33.3 per cent of the total area should be brought under forests,
which at present is only about 23 per cent. However, the pressure of agriculture, river valley projects, and new industries has delayed implementation of
such plans. For detailed descriptions of the different types of nature
reserves envisaged, the reader should refer to Ghosh and Kaul (1977).

Biosphere Reserves, National Parks and Sanctuaries are efforts to preserve
the core areas of the Himalayan landscape. Specialization in conservation has
been very useful but what is really needed is an environmental approach and
the multidisciplinary strategy of the resource scientist and the planner to see
that each environment is studied and planned for from many viewpoints
before major schemes are launched. It is desirable that living collections of
endangered and spectacular flora be introduced into botanical gardens, and
nature reserves be given protection in their natural habitat by imposing a
ban on their removal. Except for the Lloyd Botanic Garden in Darjiling,
there is no other Botanical Garden in the eastern Himalaya. It is necessary
to extend this important Garden and have additional ones in Sikkim and
Arunachal, specializing in the flora of their area. Keshab Pradhan, Chief
Conservator of Forests, Sikkim, has demarcated spectacular rhododendron
sites having the maximum variety along with alpine flora; these are the first
High Altitude Botanical Gardens in India.

Our armed forces are present in the Himalaya in large numbers. Short
courses aided by colour films and documentaries on conservation and natural
history should be run at the Indian Military Academy and other Defence
Schools of instruction so that the importance of conservation filters down to
the *jawan,* and he takes pride in preserving our heritage of forests and wild life.

Arunachal and Sikkim are happily better off than other states in protecting
wild life because of 'inner line' restrictions. These should be strictly followed
to keep the known poachers out. Sikkim has proposed the setting up of the
Kanchenjunga National Park, and Arunachal, a High Altitude National Park
around the Tawang–Sela area which abounds in rhododendrons and alpines.

Orchid sanctuaries and parks should be developed in areas where orchids
are endemic and occur in profusion. Sikkim, Arunachal and Bengal are going
ahead in setting up centres for cultivation of orchids and have already made
a very good start (Varmah and Sahni, 1970).

CONCLUSION

Raymond Dasmann (1972) has defined environmental conservation as 'the rational use of the environment to provide a high quality of living for mankind'. Since the tempo of development in the Himalaya has to be kept up and priority has to be given to projects which help to raise living standards of the people by resorting to advanced technology, the only realistic approach to the problem of the environment lies in *setting apart sizable natural areas* in representative habitats, to preserve some of our ecosystems in pristine condition as bench marks for conservation of the environment.

BIBLIOGRAPHY

ANONYMOUS. 1973. 'Materials for the Flora of Bhutan'. *Rec. Bot. Surv. India.* Calcutta. 20(2): 1–278.

BISWAS, K. 1941. 'The Flora of the Aka Hills'. *Indian For. Rec.* (NS), Delhi. (Bot) 3(1): 1–62.

BOR, N. L. 1938. 'A Sketch of the Vegetation of the Aka Hills, Assam, a Synecological Study'. *Indian For. Rec.* (NS) (Bot), Delhi. 1(4): 103–221.

———. 1953. *Manual of Indian Forest Botany*. Oxford University Press: Bombay.

BURKILL, I. H. 1924. 'The Botany of the Abor Expedition'. *Rec. Bot. Surv. India.* Calcutta. 10(1): 1–154 and 10(2): 155–420.

CHAMPION, H. G. and S. K. SETH. 1968. *A Revised Survey of Forest Types of India*. Manager of Publications, Delhi.

CHATTERJEE, D. 1939. 'Studies on the Endemic Flora of India and Burma'. *J. Roy. Asiat. Soc. Beng.* Calcutta. (Sc) 5(3): 19–67.

DASMANN, R. F. 1972. *Environmental Conservation*. John Wiley: New York.

DEB, D. B. and K. C. MULLIK. 1968. 'A Contribution to the Flora of Bhutan'. *Bull. Bot. Soc. Beng.* Calcutta. 22(2): 169–217.

DEB, D. B. and R. M. DUTTA. 1971.'Contribution to the Flora of Tirap Frontier Division'. *J. Bombay Nat. Hist. Soc.* 68:573–95.

GHOSH, R. C. and O. N. KAUL. 1977. 'Nature Reserves in India'. *Indian For.* 103(8): 497–512.

GRIFFITH, W. 1847. *Journals of Travels in Assam, Burma, Bhutan, Afghanistan & the Neighbouring Countries*. Calcutta.

GUPTA, R. K. 1972. 'Boreal and Arcto-Alpine Elements in the Flora of Western Himalaya'. *Vegetatio*. The Hague. 24(1–3): 159–175.

HARA, HIROSHI. 1963. *Spring Flora of Sikkim Himalaya*. Hoikusha: Osaka.

———. 1966. *The Flora of Eastern Himalaya*. University of Tokyo Press: Tokyo.

———. 1971. *The Flora of Eastern Himalaya*. II Report. University of Tokyo Press: Tokyo.

HOOKER, J.D. 1849–1851. *The Rhododendrons of Sikkim Himalaya*. L. Reeve & Co: London.

HUTCHINSON, J. 1969. *Evolution and Phylogeny of Flowering Plants*. Academic Press: London & New York.

JHA, M. N. 1977. A note on shifting cultivation. (Personal communication).

KINGDON-WARD, F. 1960. *Pilgrimage of Plants*. George G. Harrap: London.

LAWRENCE, G. H. M. 1960. *Taxonomy of Vascular Plants*. Macmillan: New York.

LEE, SHUN-CHING. 1935. *Forest Botany of China*. Commercial Press: Shanghai.

MERRILL, E. D. 1945. *Plant Life of the Pacific World*. Macmillan: New York.

RAIZADA, M. B. and K. C. SAHNI. 1960. 'Living Indian Gymnosperms'. *Indian For. Rec.* (Botany). 5(2): 73–150.

Rao, A. S. 1974. 'The Vegetation and Phytogeography of Assam-Burma'. In: M. S. Mani, *Ecology and Biogeography in India*. W. Junk: The Hague, pp. 204–46.

Rao, R. S. and G. Panigrahi. 1961. 'Distribution of Vegetation Types and their Dominant Species in Eastern India'. *J. Indian Bot. Soc.* 40:274–85.

Rau, M. A. 1974. 'Vegetation and Phytogeography of the Himalaya'. In: M. S. Mani, *Ecology and Biogeography in India*, 247–80.

Richards, P. W. 1952. *Tropical Rain Forest: an Ecological Study*. Cambridge University Press: England.

Ridley, H. N. 1922–25. *The Flora of the Malay Peninsula*. 5 vols. L. Reeve: London.

Sahni, Birbal. 1933. 'The Wood Anatomy of the Homoxylous Dicotyledon, Tetracentron Sinense'. *Proc. 20th Ind. Sci. Cong. Patna*, 317.

——. 1939. 'The Eastward Opening of the Himalayan Geosyncline into the Pacific Ocean'. *Proc. 6th Pacific Sci. Cong*, 241–4.

Sahni, K. C. 1960. 'Botanical Exploration in Sikkim for Fodders for Yak and Sheep'. Report submitted to Govt. of India, 1–13. Restricted.

——. 1962. 'A Conifer New to the Flora of India'. *Indian For.* 88(10): 748–9.

——. 1969. A Contribution to the Flora of Kameng and Subansiri Districts, NEFA'. *Indian For.* 95(5): 330–52.

——. 1970. 'Protection of Rare and Endangered Plants in the Indian Flora'. *Proc. IUCN 11th Technical Meeting*. Morges, Switzerland. 11: 95–102.

——. 1973. 'Protection of Endemic & Relict Taxa in Indian Flora'. Dehra Dun. *Proc. 12th Forestry Conf.*

Spate, O. H. K. 1954. *India and Pakistan: A General and Regional Geography*. Methuen: London.

Stearn, W. T. 1960. '*Allium* and *Milula* in the Central and Eastern Himalaya'. *Bull. Brit. Mus. (Nat. Hist.) London.* 2:161–91.

Steenis, C. G. G. J. Van. 1948–58. *Flora Malesiana*. Noordhoff-Kolff N. V: Djakarta.

Twenty-Five Years of Forestry in Arunachal. n.d. Director of Information and Public Relations, Arunachal Pradesh.

UNESCO. 1974. *Task force on criteria and guidelines for the choice and establishment* of *biosphere reserves*. MAB Final Report No. 22. UNESCO, Paris.

Varmah, J. C. and K. C. Sahni. 1976. 'Rare Orchids of the North Eastern Region and their Conservation'. *Indian For.* 102(7): 424–31.

Takhtajan, A. 1969. *Flowering Plants, Origin and Dispersal* (Tr. Jeffrey) Oliver & Boyd: Edinburgh.

CHAPTER 3

HIMALAYAN FAUNA

M. K. RANJITSINH

The climate, terrain and low atmospheric pressure obtaining at high altitudes has obviously required certain physiological development within animals for them to adapt successfully to such an environment. Such adaptation can be broadly classified into two types—the evolution of certain specialized members of some animal families, notably *Cervidae, Canidae* and *Felidae;* and the proliferation of those genera already adapted to mountain habitats—the *Capra, Ovis, Capricornis* and others—into a number of species and subspecies.

In fact, however, relatively very few of the larger members of the animal kingdom have been able to adapt themselves to a high-montane topography, and vast tracts of the world's mountain ranges are almost devoid of large mammalian life. This is particularly so with the mountains of continents other than Asia; the most noteworthy example being the Andes of South America, the world's largest single mountain-chain. It is remarkable therefore, that the Himalaya, the world's highest and one of the steepest and youngest mountain ranges, should nonetheless possess almost a third of the world's larger mammalian species that could be called true mountain animals.

It would not be possible in this short space to discuss all the larger animals, let alone Himalayan fauna, in their entirety. The more prominent mountain species and the endangered ones, therefore, will have to suffice. A list of the major animals found in the Himalaya and their foot-hills is given in the Appendix.

Ranging from the areas of permanent snow and the vast upland plateaux, bleak, dry, windy and bitterly cold, to the temperate-montane middle zone and the lush subtropical foot-hills, barely a few feet above sea level, the Himalaya encompass a diversity of habitat almost unequalled in the world. Included in these zones are subtropical pine forests, montane wet-temperate, moist temperate, dry temperate, subalpine and alpine forests. Above the tree line are the stunted rhododendron and juniper thickets beloved of the musk deer, and the meadows reaching up to the permanent snow line.

It would be appropriate to begin with the wild goat, animals which epito-mise the Himalayan crag more than any other. The Markhor, the most spectacular of them all, is found in the Himalaya from the Pir Panjal range in Kashmir to west of the Banihal Pass, to the Hindu Kush and Karakoram Ranges. The taxonomy of the animal has been a subject of much controversy. The straight deep-corkscrew type of horn formation found in the Pir Panjal, Kaj-i-Nag, Kaffir Khan and Shamsabari Ranges is referred to as the Pir Panjal race or subspecies, while the out-flaring horns of the animals found in Astor and Haramosh have earned them the name of the Astor race. Further north in Chitral, the horn formation again resembles the Pir Panjal type. A third race the Kabul Markhor, occurs further west, and has a narrower spiral. There is however no cast-iron rule as to the horn shape in the areas described. Not only are intermediate shapes found, but speci-mens carrying Astor and Pir Panjal horn types have also been seen in the same herd.

Not endowed with the underwool of the Ibex, the Markhor prefers alti-tudes below the snow line. In most parts of its range in southern Kashmir it prefers forested slopes, emerging into the grassy meadows to graze. Further to the north and west it is well adapted to open crags, but nowhere is it averse to entering the timber as the Ibex is in the Himalaya. Grazing is done mostly in the mornings and evenings as with most hoofed animals in the mountains, the afternoon being spent usually on a commanding and often unassailable rock buttress or cliff. No animal with the exception of the Tahr invariably chooses so difficult a retreat to rest in, and the presence of cliffs to escape into is a necessary concomitant of a Markhor habitat. An hour-long sight of a Markhor buck, statuesque on a pinnacle of rock gazing into the stupendous chasm below him, fifty-inch horns glinting in a setting sun and only the swaying of his flowing dirty-white hair, reaching to his knees, revealing him to be an animate being, is perhaps the grandest spectacle that the author has been privileged to witness.

Accounts of sportsmen testify to the impressive number of Markhor in the past. Despite the inaccessibility of the terrain it inhabits, the animal has been drastically reduced due to illegal hunting, partly by the armed forces but mainly by local inhabitants, who have the endurance to pursue the animal into the most remote fastnesses. A survey carried out by the author in the Kaj-i-Nag range to the east of the cease-fire line in 1969 re-vealed less than 200 Markhor.

The males separate from the nannies after the rut (mating periods), though some younger bucks, called *reend* in Kashmir, often continue to remain with the females. The largest all-male group seen by the author is of six, and the largest mixed herd of fourteen specimens. Formerly, herds of up to sixty animals have been recorded.

The Ibex is the second member of the goat family, found in the Himalaya west of the Satluj gorge. Of about the same height as the Markhor but less squat, with sweeping horns curving backwards measuring up to 125 cm or more in western Ladakh and Chitral, the Ibex occupies the highest altitudes of all goats in these mountains. Gifted with a fine sense of smell, and with excellent vision, the Ibex is at home on the cliffs as well as in the upland pastures. The midday siesta is at times deliberately spent on snow fields—partly to escape the flies in summer and partly because of better visibility all around.

It is sad that despite its vast range this fine animal has also been severely decimated. Search in the Kulu valley resulted in its being sighted only in the Solung and Hamta Nalas at the head of the valley, and a fortnight's wanderings in Ibex country in the Kargil area of Ladakh on both sides of the Indus, resulted in only two groups being seen in two nalas.

The same fate has been suffered by the Himalayan Tahr, a shaggy, heavily-built goat with short curving horns. Occupying the Himalayan ranges from the Pir Panjal and Kishtwar in Kashmir to Bhutan in the east, the Tahr has the place occupied by the Markhor further westwards. However, the Tahr is even more a creature of the forest than the Markhor, almost never going above the tree line, and even more loth to leave the cliffs. In New Zealand, however, it has adapted itself to open hill sides and grassy slopes.

It is difficult to imagine that an animal inhabiting such steep and forested terrain could be so reduced by hunting; in fact, their reluctance to change their habitat has contributed to their decline. It is a sad commentary that there are perhaps more Tahr in New Zealand—where they were introduced in the beginning of this century and are regarded as pests—than in the whole of the Himalaya.

Of the *Rupicaprids* or goat-antelopes, the largest and most elusive member of this genus is the Takin. A large heavily-built animal resembling the American bison in appearance and gait, chocolate-brown in colour with a golden saddle on the back, the animal is found in northern and north-eastern Bhutan; the Mishmi Hills; and perhaps the headwaters of the Subansiri River—the last two in Arunachal Pradesh. However, fossils of what was obviously the progenitor of the Takin have been found in the Siwalik Range of the Himalaya far to the west. Gregarious by nature, the Takin travel long distances between their summer and winter habitats, and are at home in forest as well as in rhododendrons. The author has observed the Takin only in northern Bhutan, the most rewarding occasion being two days spent with a herd of about 80 animals on the Bhutan-Tibet border in 1965.

The other two members of the genus, the Serow and the Goral, have short, sharp, curving horns; the former is the larger animal. Both have two races each in the Himalaya, the animals in the eastern Himalaya being more brown and russet. The agile Goral prefers grassy slopes with broken ground, usually at lower altitudes in the southern Himalaya, though the author has encountered them as high as over 3050 m. The Serow is a recluse, solitary and secretive by nature, occupying small cliffs interspersed with thickets. In all the areas where the author has encountered them—Kashmir, Bhutan and Thailand—they have very restricted movements in their selected habitats.

What is true of the goats also applies to the wild sheep, for no mountain range of the world boasts of more species of these than the Himalayan uplands. Pride of place must go to the Marco Polo sheep with a large ram's horns making a complete circle and a half. The record pair measures 187.5 cm exceeding by far the ratio of horn-length to body-height of any animal in the world. Only an intermittent visitor to Hunza in Pakistan, it is hoped that the formation of the new Khunjerab National Park will succeed in establishing a sizable herd permanently in this area.

Smaller in horn length but with a larger body, the *Ovis Ammon Hodgsoni* or Nayan is the king of the Tibetan and Trans-Himalayan plateaux. Crossing the main Himalayan divide only at a very few places and irregularly, in Nepal and Bhutan, and in Spiti in India, its main abode in the Himalaya is to the east of the Zaskar range in Ladakh. Occupying bare rolling mountains of 4270 m and more—intensely cold and windy terrain, the Nayan are preyed upon by wolves. The Nayan does not appear to be a very gregarious animal; the largest mixed herd seen by the author had only 14 members, and the largest all-male group 8, led by two patriarchal rams with white necks and chests. As with most animals of the high uplands, the Nayan moult in summer, the underwool peeling off to give the animal a ragged look.

Slightly lower than the Ammon and frequenting broken ground are the Bharal or Blue Sheep—which in fact are not true sheep! Gray-blue brown in colour, the mature ram develops a black neck and throat patch and almost cylindrical horns growing outwards and then backwards.

Occupying Himalayan and Trans-Himalayan zones, the Bharal is found from Zaskar in Ladakh eastwards to the Brahmaputra Gorge. However, it does not occur everywhere to the south of the main Himalayan divide, and Ladakh and Bhutan are the areas in this range where it is the most numerous. In Ladakh the Bharal have been severely decimated, and the herds which were so common in 1958 were not encountered anywhere by the author on a subsequent trip in 1970. In Bhutan, however, they are still relatively plentiful and were met with in substantial numbers both in 1965 and in 1974.

Another kind of sheep which has been over-hunted for its meat is the Urial or Shapu, found in Ladakh a little east of Leh to Astor, Chitral and Gilgit, and beyond in the west. Usually occupying the lower hill-ranges in the main river valleys, the Shapu were the first animals to suffer from the presence of the armed forces. In western Ladakh where they were numerous in 1958, in 1970 they were found in only three areas, with a herd of 62 animals being the only sizable group encountered. The black-and-white throat patch, with slender horns forming a circle makes the adult Shapu ram one of the handsomest animals in the Himalaya. When alarmed they emit a shrill whistle.

Sambar and the barking deer are found high up in the Himalaya in the southern ranges, but they are not true mountain animals. But two other grand species of the deer family—the Hangul or Kashmir Stag, and the Shou or so-called Sikkim Stag inhabit the Himalaya—though they have never been recorded in Sikkim. They are also amongst the rarest animals in the world. The Hangul, also called Barasingh because of the twelve or more point antlers it carries, was found till very recently in the northern valley of Kashmir and the valleys which lead into it—Lidder, Sing, Lolab and others. Eastwards it occurred in Wardhawan, and Kishtwar upto Chamba, migrating each summer to higher altitudes like Haramukh and Gurais. The number was perhaps never very large—one estimate being of 5000 animals sixty years ago. The Hangul has been pursued by hunters as a trophy and even more so by poachers for its meat and skin during the rutting season, when the animals came down to the lower slopes and the stags revealed their presence by their hoarse roaring calls. As a result, the Barasingh has been reduced to a total number of not many more than 200, most of them being in Dachigam Sanctuary—their only real chance of survival.

The future of the Shou in the Himalayan range appears to be equally grim. It is no longer found in Bhutan, and its existence in the Chumbi Valley appears to be threatened by the military roads built there by the Chinese. Further east possible sightings have been reported in parts of northern Arunachal Pradesh.

The deer that is perhaps best known is the Musk-Deer. Inhabiting dense thickets and forests interspersed with rocky ground, often at heights of 4270 m, the Musk-Deer possesses a greyish-brown coat of short, brittle, thick, insulated hair, disproportionately long legs, a pair of downward and backward curving tusks instead of horns in the males, and what is even more important, a musk pod which has proved to be its nemesis. This musk gland produces the legendary *Kasturi*, which gives off a strong odour when dry, and commanding an exhorbitant price as an ingredient for perfumes and medicines, has made the animal the chief target of every Himalayan poacher.

There are other animals which cross over the main Himalayan divide at a few places—occasionally and in small numbers—and which are truly the animals of the Trans-Himalayan ranges in Ladakh. The Wild Yak, the huge shaggy black bovid—the largest wild animal in the Himalaya —is now only a visitor to Changochenmo. The Tibetan Antelope or Chiru with its almost straight long horns, a specially adapted breathing-system and the finest underwool to withstand the extreme cold, does visit eastern and northern Ladakh where the army jawans take a fairly heavy toll of them. The Goa or Tibetan Gazelle, formerly found in small numbers in the grassy valleys have all but disappeared in Ladakh. Though seen by the author at Tso Kar in 1958, an intensive search for them in the Hanle area revealed only two pairs of footprints in 1970. In Bhutan, however, they do survive in at least one place.

To those who have met with the inquisitive and ubiquitous Kyang in Ladakh and elsewhere, it would be difficult to believe that any harm could befall this very handsome Wild Ass, which has white underparts offsetting a deep copper-russet coat. But there is no limit to man's desires and even the Wild Ass is shot as a trophy or for meat.

The proliferation of the larger ungulate species mentioned above, together with the array of smaller mammals and birds that inhabit the Himalaya, is matched by the evolution of a very impressive number of species in the order Carnivora.

In the *Felidae* family, both the tiger and leopard, especially the latter, have been observed deep in the Himalaya and at high altitudes. A tiger has been recorded at 3050 m in Garhwal, and the range of the panther exceeds that of the Snow Leopard in Kashmir and many places elsewhere. The Clouded Leopard occurs in the southern ranges of the eastern Himalaya, as do a number of smaller cats. However, in view of the fact that the other animals are not special mountain creatures, the discussion will be restricted to the Snow Leopard and Lynx.

Endowed with a magnificently speckled, fluffy pelt, large paws and an exceptionally long and bushy tail, the Snow Leopard is the main predator of the higher Himalaya. Its range is immense—covering in fact the entire Himalaya, extending from altitudes as low as 1830 m to 5185 m or more in summer. It is the chief natural predator of the Ibex, Markhor, Bharal, and Shapu, and takes toll of all the large ungulates of the uplands and crags. Though known to prey on domestic livestock, it rarely becomes a sheep-and-goat-lifter after the fashion of the leopard. It is primarily dependent upon wild animals for its food and the reduction of the wild ungulate population has had a marked impact upon the Snow Leopard population, which is further accentuated by the greed of the trapper and poacher for the prices its pelt commands. Never very numerous, the Snow Leopard has become a rarity in most parts, with individuals having to cover huge territories to obtain sustenance. In a census carried out by George Schaller in Chitral, about 3000 square kilometres were found to hold only four or five specimens. Bhutan and Ladakh, luckily, still have a sizable number of Snow Leopards.

The second-largest cat of the higher altitudes is the Lynx, with the short tail and large black-tipped ears characteristic of the species. Though more diurnal in its habits than is normally believed, its small size, camouflage, and the extreme altitudes it inhabits together with its shy nature, preclude frequent encounters. Though it preys largely on hares, marmots and birds, it is known to kill young bharal as well. Its costly skin makes it greatly prized. Ladakh again is the home of the largest number of this animal.

Another felid of the high mountains is the beautiful Pallas' Cat of which the author was fortunate enough to obtain a fleeting glimpse in Ladakh. Very little is known about the habits of the animal.

As regards the *Canidae*, though the wild dog does occur in the Himalaya at altitudes of as high as 2440 m or more, the reduction of its prey species has affected its numbers greatly and it is now very rarely to be met with anywhere in these mountains. The largest and most important member of the family is the Tibetan Wolf, which is the chief predator of the Trans-Himalayan uplands and plateaux. The author has never seen them in numbers larger than two, but packs of up to 20 have been seen as well as some black specimens, which are only aberrations in colour and do not form a separate race. They feed upon hares and marmots, as well as upon Ovis Ammon, Goa, Chiru, Shapu, and whenever the slopes are not too precipitous, the Bharal. The largest toll is taken in spring when the young are born, and in winter when their padded feet give them an advantage over hoofed animals in the snow. The skulls that are seen on high passes and at *chortens* in Ladakh bear testimony to this prowess. It is almost the same animal that is found in central Asia, northern Europe and America—testifying to its abilities to colonize and adapt.

The bear family is represented in the Himalaya by the Brown and the Himalayan Black Bear. A third member, the Tibetan Blue Bear, is so rare that practically nothing is known about it. One was reported across the border into Tibet when the author was camping on the Bhutan–Tibet border in 1965. Further to the south-east of the Himalaya the Malayan Sun Bear appears, but it is not a Himalayan animal.

The Brown Bear, also referred to as the Red or Snow Bear, is the largest carnivore in the Himalaya. It is omnivorous, and stores immense quantities of fat for its winter hibernation period. It is not averse to raiding flocks of domestic sheep and goats, and assiduously hunts rodents, while carrion from animals which have perished in the winter provides a welcome change of diet. Though known to raid crops and fruit trees, its remote habitat and shy nature makes it less of a nuisance to man in this regard than the Black Bear. Inhabiting the Himalaya west of the headwaters of the Ganga, the Brown Bear is found at the tree line and above. In Chamba, Kishtwar and Saru it is still quite numerous, though to the west in Pakistan the numbers have declined drastically. Like all members of the ursus family it is devoted to its young—which in this case are born during hibernation—and the cubs spend a long time under parental tutelage. In the Kaj-i-Nag mountains in Kashmir in 1969, the author watched a large sow Brown Bear resting in the afternoon while her cub ventured out on forays against its mother's wishes. Five times she had to bestir herself and bring back the errant cub. Calls and even cuffing was of no avail. Ultimately she half lay upon the cub, one huge arm pinning it to the ground, while it howled in protest.

After a while she let it go, whereupon, thoroughly subdued, it curled up beside its mother and went to sleep.

No animal in the Himalaya has a larger range and no animal is known and feared more than the Himalayan Black Bear. Extending along the whole range and further east into south-east Asia, this bear is to be met with from the lowest Himalayan foot-hills to the snow line. In November 1959 the author saw a large male is Solung Nala of Kulu at an altitude of 418 m with freshly fallen snow all round. Being good climbers and catholic in their diet they have been able to make the fullest use of their habitat, and human habitation and cultivation is often an attraction rather than hindrance. The skins and meat are of no value, and were it not for its propensity to raid crops, the animal would have no enemies. As it is, it is the safest of the larger Himalayan animals from human attack.

The Lesser or Red Panda, though a dimunitive animal barely 91.5 cm long and weighing 4.5 kgm, is one of the best-known of the smaller Himalayan animals. With pointed face striped with black and white, surmounted by large white-rimmed ears, the body marked in deep rusty-red and blackish underparts, the Red Panda is one of the most beautiful animals in existence. It is easily tamed and is one of the greatest prizes for the illegal animal dealer. Being largely arboreal and nocturnal in habit, it is seldom seen. Mainly a vegetarian, its range extends east from the Nepal Himalaya at altitudes between 1830 to 3660 m. Its nearest relative is the famous Giant Panda of Szechwan province in China.

No chapter on Himalayan fauna would be complete without a mention of the legendary Yeti or Abominable Snowman. Known by various names to the local inhabitants of the higher Himalaya almost throughout the range, the Yeti first came into prominence in 1951 when pictures of foot-prints of a large unidentified bi-pedal taken near Everest were published by Eric Shipton. Since then a number of similar large footprints, 30 cm or more long, have been recorded, some at a height of 7015 m. In 1970, a member of the Annapurna expedition in Nepal not only took photographs of such footprints but saw a hairy animal in the distance. Voluminous literature exists on the subject and special expeditions have been made to obtain a specimen, or at least conclusive proof, of its existence. The missions have not succeeded, and the conical scalp believed to be of the Yeti in Thyangboche near Everest, turned out to be the skin of a Serow. No satisfactory explanation has yet been given of the footprints which resemble none of the known animals—bears, Himalayan Langur or Snow Leopard. Thus the mystery and the romance of the Yeti continues, to lend yet more magic to these supreme mountains.

It is evident from the above that as in most of southern Asia, wild life is in sharp retreat almost throughout the Himalaya. But, in the Himalaya, unlike the rest of this region, the destruction of habitats and competition from domestic stock has contributed comparatively little to the decline.

It is undoubtedly true that the alpine and subalpine forests have been over-exploited and a tremendous amount of soil erosion has taken place. But with the possible exception of the Hangul, habitat destruction has not had a marked effect on the larger mammals. The reasons for this are manifold. The forest-dwelling animals of the higher Himalaya are not dependent upon dense and extensive forests. In the lower Himalayan reaches the increase in the number of buffaloes owned by Gujar herdsmen, which move upwards every summer, has had the most damaging effect on the biotope. But in the upper meadows, domestic sheep and goats have not caused such havoc, partly because climatic and topographical factors ensure that they cannot inhabit these meadows all year round, but mainly because the number of wild ungulates is so low even in the best of times that there is no serious competition for food. Only in spring is there any real overlapping of the territory of wild and domestic ungulates—when the receding snow results in fresh grass which attracts both the winter-hungry wild goats and sheep and the upward-moving domestic stock. But even here there is still plenty of place for the wild animals. Further, changes in the topography—soil erosion, for instance, altering a grassy or forested hill-side into shale and rocks—may make the habitat unsuitable for animals such as deer, but the area may quickly be taken over by others such as wild goats and the *Rupicaprids*.

No, the main blame for the reduction of the larger wild mammals must be directly and indirectly attributed to the poacher and the trapper. Herdsmen in the summer, and hill-dwellers in the winter when snow makes tracking easy and puts the hoofed animals at a disadvantage, have always taken a toll, often with illegal firearms. But in the last quarter-century, arms licences have been freely given to the local people, and army personnel have been known to shoot up to fifty sheep in a day, sometimes from the air. What is even more dangerous is arming the nimble hillmen, usually employees, to go and obtain meat for the army pickets. And all the while rapidly-increasing prices of furs and musk have made what were previously occasional furtive forays for meat, into a full-time occupation for the local inhabitants. No animal population can withstand repeated commercial overkill, least of all those from the less prolific temperate zone.

For the magnificent fauna of the Himalaya to be saved from total extinction in some parts, severe curbs must be placed on the sale of certain types of fur, and a complete ban on trade in some others, as well as of musk. Illegal hunting both by the army and civilians must be halted, and arms licences cancelled, and large, viable national parks and sanctuaries established to protect not only the endangered species but the total range of representative fauna, flora and habitats in the Himalaya in the different zones. In other words, it would not suffice to set up a single sanctuary for one or more species, though this in itself is also important. The same species may be found in different geomorphological habitats which may make it different in habits, behaviour and even in physiology. Besides, different geomorphological formations and the flora they support are worth preserving in themselves.

The prevention of human exploitation, particularly of stock grazing, is difficult, and it is for this reason that so few national parks have been established in the Himalaya. In certain places, however, exploitation will have to be stopped—not only for the sake of the animals but for safeguarding the watershed as well. In places where this is absolutely impossible or not really necessary, wildlife sanctuaries will have to be established, but great control would have to be exercised to ensure that the partial human use permitted is not exceeded.

Captive breeding should be the last resort in attempting to save a species from extinction, but it is less appropriate for Himalayan fauna than for the plains species. Such programmes, to be successful, should preferably have sizable numbers of one herd as a unit in captivity. It is usually extremely difficult to get even an individual specimen let alone a whole herd of mountain animals. In the case of rare species—the Hangul for instance—the problem is much more severe, and the question presents itself as to whether it would be wise to capture a herd of such rare animals. Even if animals are bred in captivity the problems of restoring them to the wild would be very great.

The experiment of raising musk deer and extracting the musk periodically from the pod for commercial purposes without killing the animal, has apparently been successful in China. But even if such an endeavour were to succeed in the Himalaya, the question is, will this save the species in the wild? The market for musk, now partially met by synthetic musk, is insatiable. Thus, even if substantial amounts of real musk are produced commercially—a difficult task by any account—the price of musk will not be sufficiently affected to be a disincentive to the poacher. Therefore, for the animal to be saved, it must be protected in the mountains, in parks and sanctuaries.

A brief survey of existing national parks and reserves and those that need to be established, may not be out of place. In Pakistan, perhaps the most significant step has been the establishment of the Khunjerab National Park in Hunza in 1975 by special order of the President. With an area of 2200 square km, it is the only refuge in the Himalaya for that most spectacular of all sheep—the Marco Polo. Bharal and Snow Leopard are also present. The Chitral Gol sanctuary preserves a reduced herd of over a hundred Markhor. More sanctuaries are needed for the Markhor and Ibex, and especially for the endangered Shapu.

In Kashmir the most important sanctuary is Dachigam. Not only is it the only retreat for the Hangul, it holds both Brown and Black Bear, Serow, Snow and ordinary Leopard, Musk Deer and wild boar. Proper sanctuaries need to be established in Kaj-i-Nag and Shamsabari ranges for Markhor, bear and Musk Deer, and in the Lidder–Wardwan–Kishtwar divides and in Bhadarwa for the remnant Hangul, Ibex and Tahr, amongst others. In crucially-important Ladakh, at least one park or sanctuary needs to be established in each of five regions—the Kargil-Lamayuru area for Shapu and Ibex; Zaskar mainly for Bharal; the huge Changthang area of Hanle and the inland lakes Tso Kar and Tso Morari for Ovis Ammon, Bharal, Snow Leopard, Lynx, Tibetan Gazelle, Wild Ass and Wolf—and the rare Black-necked Crane; the upper Shyok for Ibex, carnivora and the Tibetan Antelope; and the western Changehenmo for the Antelope and Wild Yak.

Himachal Pradesh has about thirty sanctuaries and reserves. Most of them, however, are in the lower ranges and almost all are very small and not viable. An important reserve is Gamgul on the Chamba–Kashmir border where Hangul are reported to migrate in winter. This needs to be effectively protected, and reserves need to be established on the Chamba-Bara-Banghal divides for Ibex, Bharal and the carnivora. Establishment of a national park north of Manali in the Kulu valley has long been under consideration and is an urgent need. Apart from its scenic beauty, it is suitable for trekking and is one of the most convenient areas for the viewing of Himalayan fauna. The Kanawar–Parbati Valley divide is another area that requires conservation.

The state of Uttar Pradesh has three important sanctuaries in the Himalaya—Nanda Devi, Kedarnath Musk Deer Sanctuary, and the Gobind Pashu Vihar. The last is the most important, holds bharal and other high-montane fauna, and needs to be declared a national park. Another area of significance is the Rari-ka-danda on the Yamuna–Bhagirathi divide.

In Nepal three national parks are being established. The Sagarmatha National Park encompassing Mt Everest, and the Langtang National Park, are true high-altitude parks ranging from mixed broad-leaved forests to perpetual snow, and are increasingly popular for trekking; this in turn brings problems of environmental despoilation in its wake. The third park, at Lake Rara, Nepal's largest lake, is on the flyway of migratory birds. The human population may be re-settled elsewhere from Lake Rara, but in the other two parks the original inhabitants, the Sherpas and their livestock, will remain part of the ecology of these mountains.

These three areas do cover most of the high-montane fauna of Nepal including the Red Panda, with the exception of the Bharal, which is being protected in the Shey Reserve. However, more parks and reserves would be needed to protect some of the remaining strongholds of the Bharal and the Tahr. Nepal is not as rich in the larger mammalian mountain fauna as the western and eastern Himalaya, partly because the extreme altitudes of the main Himalayan divide here largely prevents influx of Trans-Himalayan fauna.

In north-western Sikkim, Wild Ass have been reported, and some Bharal seen at the base of Mt Kanchenjunga. Tahr are also found sporadically.

The Lachung Valley in northern Sikkim is an important area for conservation.

Bhutan is not only one of the last great strongholds of Himalayan fauna, but offers perhaps the greatest hope of their continued survival. The three contiguous sanctuaries—Laya, Gasa and Jigme Dorji—covering more than 7770 sq. km, the entire northern frontier of the country—constitute the largest officially-protected area in the entire Himalaya. It contains the largest population of Bharal, Takin, Musk Deer and Snow Leopard in the whole range. In the middle Bhutan zone, however, there is at present only one small National Park, mainly for Goral, which in this area is very numerous. More sanctuaries need to be established in this zone for the other Rupicaprids, deer and carnivora.

The Bhutanese being Buddhists, are traditionally not hunters. Hunting has however started, particularly for musk, and could very well become excessive if not controlled soon.

Further east in the state of Arunachal Pradesh in India, the only sanctuary of note at present is Namdapha. The large tribal populations with their proclivity for hunting have denuded vast tracts of most of the larger animals. The ecology and fauna of this region are important, however, and the upper reaches of the Subansiri and Tsangpo Rivers and other remote valleys in the north need to be surveyed and effective sanctuaries established. An area in crucial need of conservation is the Mishmi Hills, one of the very few habitats of the Takin, which is severely endangered in these parts.

The outlook for wildlife in most parts of the Himalaya, therefore, is grim; in some places even desperate. Because of the climatic conditions and inaccessibility, the problems encountered in conserving fauna of the plains do not exist in the mountains—mainly because the human pressure on land is not as great and the poachers are relatively few, and well known. Besides, furs have to be sold through dealers, and if the sale and export of fur were rigidly controlled, poaching would not be worthwhile. What is required is a relatively small outlay on manpower of the right type—and the political will to establish parks and reserves, and implement the laws. On this will ultimately depend whether these grand mountains continue to be the great repository of the larger mammalian mountain fauna of the world, or become totally devoid of it.

APPENDIX

Larger Mammals Found in the Himalaya

1. Snow Leopard (*Panthera uncia*)
2. Tiger (*Panthera tigris*)
3. Leopard or Panther (*Panthera pardus*)

4. Clouded Leopard (*Neofelis nebulosa*)
5. Golden Cat (*Felis temmincki*)
6. Leopard Cat (*Felis bengalensis*)
7. Marbled Cat (*Felis marmorata*)
8. Pallas's Cat (*Felis manul*)
9. Lynx (*Felis lynx isabellinus*)
10. Binturong (*Arctictis binturong*)
11. Spotted Linsang (*Prionodon pardicolor*)
12. Tibetan Wolf (*Canis lupus*)
13. Wild Dog or Dhole (*Cuon alpinus*)
14. Red Fox (*Vulpes vulpes*)
15. Tibetan Fox (*Vulpes ferrilatus*)
16. Himalayan Brown Bear (*Ursus arctos*)
17. Himalayan Black Bear (*Selenarctos thibetanus*)
18. Lesser or Red Panda (*Ailurus fulgens*)
19. Ermine (*Mustela erminea*)
20. Weasels (*Mustela sibirica, Mustela kathia and Mustela altaica*)
21. Martens (*Martes foina intermedia, Martes flavigula*)
22. Marmots (*Marmota bobak himalayana, Marmota Caudata*)
23. Hog Badger (*Arctonyx collaris*)
24. Himalayan Crestless Porcupine (*Hystrix hodgsoni*)
25. Tibetan Wild Ass (*Equus hemionus kiang*)
26. Wild Yak (*Bos grunniens*)
27. Marco Polo Sheep (*Ovis Ammon polii*)
28. Ovis Ammon or Nyan (*Ovis ammon hodgsoni*)
29. Urial or Shapu (*Ovis vignei*) (*Ovis orientalis*)
30. Bharal (*Ovis nahura*)
31. Markhor (*Capra falconeri*)
32. Ibex (*Capra ibex*)
33. Himalayan Tahr (*Hemitragus jemlahicus*)
34. Takin or Mishmi Takin (*Budorcas taxicolor*)
35. Serow (*Capricornis sumatraensis*)
36. Gorals (*Nemorhaedus goral, Nemorhaedus hodgsoni*)
37. Tibetan Gazelle (*Procapra picticaudata*)
38. Tibetan Antelope or Chiru (*Pantholops hodgsoni*)
39. Kashmir Stag or Hangul (*Cervus elaphus hanglu*)
40. Shou (*Cervus elaphus wallichi*)
41. Musk Deer (*Moschus moschiferus*)
42. Sambar (*Cervus unicolor*)
43. Barking Deer or Muntjac (*Muntiacus muntjak*)
44. Wild Pig (*Sus scrofa*)

CHAPTER 4

EARTHQUAKES IN THE HIMALAYA

H. M. CHAUDHURY

The Himalaya extend from Nanga Parbat in the west to Namcha Barwa in the east in the form of a great arc whose western end bends towards the south-west (Wadia's syntaxis) and joins the Sulaiman Kirthar ranges, whereas the eastern ranges form a spectacular hair-pin bend in Arunachal in Assam, and continue southwards across the border states of India and then through Burma. The whole sequence of bends gives the Himalaya a striking arcuate disposition. All these arcs are turned towards the Peninsula, as if they have been thrust against it.

The longitudinal section of the Himalayan range may be classified into four broad zones running parallel to each other: the Outer Himalaya, the Lesser Himalaya, the Great Himalaya, and the Trans-Himalaya. The Outer Himalayan zone or the Siwalik zone of foot-hills is 10 to 60 km wide with altitudes rarely exceeding 1 km. The Lesser Himalaya comprise a zone 60 to 80 km wide, and have an average altitude of 3 km. They consist of parallel ranges in the Nepal and Punjab regions, but of scattered mountains in Kumaun. The Great Himalaya or Central Himalaya comprise the zone of high snow-capped peaks which are about 120 or 140 km from the edge of the plains. This zone shows both sedimentary and metamorphosed rocks. The Trans-Himalayan zone, about 40 km in width, contains the valleys of rivers rising behind the Great Himalaya. These river basins are about 4 km above sea level and consist of rocks of Tibetan facies.

Between these mountain ranges and the stable peninsular mass towards which they appear to be pushed, lies the Indo-Gangetic valley, a broad, monotonous level plain built up by recent alluvium through which many rivers flow sluggishly towards the seas.

Structurally, the Peninsula represents a stable mass which has remained quiescent, except at its coasts where there have been marine transgressions. In contrast, the Himalayan mountainous zone has recently undergone earth movements of stupendous proportions. The strata of the crust in this region are marked by complex folds, reverse faults, overthrusts and nappes of great dimension. There is evidence that these movements still continue, for the region is unstable and is frequently visited by earthquakes. The earlier movements, geologists infer, consisted of a series of great impulses each of which lasted several thousand years. This explains the alternation of high mountains and deep valleys across the ranges. Looked at in this context, the Gangetic valley is a sag in the crust, probably formed contemporaneously with the uplift of the Himalaya. It is thus in the nature of a foredeep. Due to this feature and its arcuate disposition, the Himalaya have often been compared with the island arcs of the Pacific, many of which are convex and front upon oceanic troughs or foredeeps. Many other geophysical features, such as a belt of strong gravity anomalies indicating lack of equilibrium along the front of the hills, give further evidence as to the active movements in the region. As mentioned earlier, many violent earthquakes have been located in and around this belt and confirm these geological inferences.

Earthquake Occurrence

Earthquakes in the Himalayan region, which are only manifestations of the distress of the earth's crust under tremendous geological forces, are therefore connected mostly with the birth and the subsequent growth of the Himalaya, and their occurrence must date back very far. Records of earthquake occurrence are however very limited. From all the available historical records, T. Oldham of the Geological Survey of India has compiled a catalogue of earthquakes up to 1869. Although the first earthquake listed occurred in AD 893, there are very large gaps in time and many lacunae. The catalogue did however, for the first time, help to identify earthquake-prone areas in and around India. With the development of instrumental methods of observing and recording earthquakes, more homogeneous catalogues have been available from 1916. The literature thus consists of two broad categories—that of the pre-instrumental and the post-instrumental periods.

   Although no accurate instrumental location of earthquakes was possible, the Geological Survey has contributed a number of field studies of strong earthquakes during the pre-instrumental period. From observations on the site of the effects of earthquakes, a number of very illuminating reports have been published. Through such studies, the location of the earthquake source region, the nature of forces responsible for the quake, the nature of ground movements that caused the observed damage, and the nature of the wave propagation were inferred. Much of this information is now obtained by analysis of seismic recordings at a number of observatories. Nevertheless field studies of actual damage continue to be a vital part of investigations.

   The report on the great Assam earthquake of 1897 by R.D. Oldham is a classic in seismological literature. In this volume Oldham has shown how, by the application of simple principles and physical reasoning, important inferences could be obtained from observations on the direction of the throw of loose objects, rotation of tombs and pillars, oscillation of hanging objects, disposition of ground fissures, etc. In this volume, Oldham also gave evidence for the first time of different types of earthquake waves.

The detection of earthquakes and location of their sources through instrumental recording began in India in 1898 when instruments were installed at the Calcutta, Bombay and Madras observatories of the Meteorological Department. Since then the IMD has been the official agency for earthquake studies and has been developing techniques in this direction. At present the department runs a total of thirty-eight seismological observatories (Figure 7.1) and locates even small earthquakes.

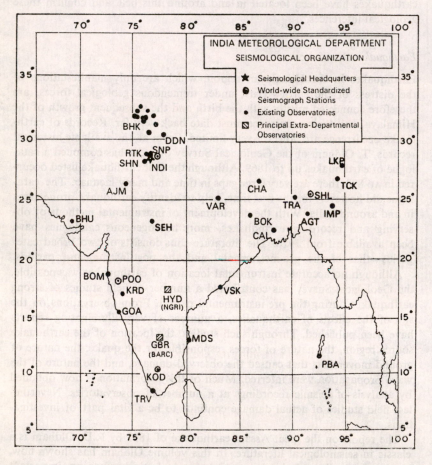

Figure 7.1 Seismological observatories in India

Figure 7.2 shows all the earthquakes in Oldham's catalogue up to 1869 and later occurrences up to 1975. Each earthquake is denoted by a circle, whose size indicates the size of the earthquake, which is a measure of the energy released and hence of the damage potential. It is clear from this map that

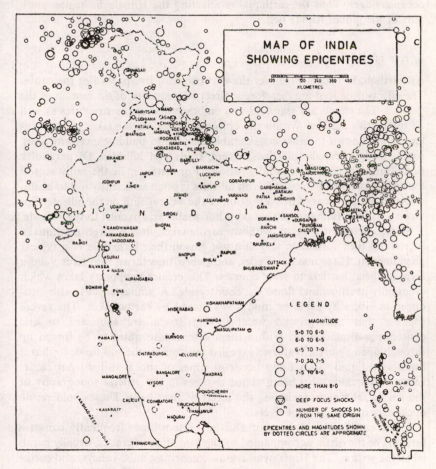

Figure 7.2   Earthquake epicentres in India

earthquakes have occurred all along the Himalayan region and the Rann of Kachchh. The Andaman-Nicobar Islands area is only a continuation of the Manipur-Burma earthquake belt. On the whole, no large earthquake has occurred in the Peninsula. Earthquake occurrence in particular areas is not exclusive to the Indian region, but throughout the world, it is mostly restricted to narrow 'seismic' belts. The earthquakes are mostly confined to (1) the circum-Pacific belt which passes through the west coasts of South and

North America, Alaska, Kamchatka, Japan, the Philippines and the Solomon Islands region, and (2) the Alpide belt through Java, Sumatra, the Andaman-Nicobar Islands, Burma, the Himalayan region from Assam to Kashmir, Afghanistan, Pakistan, Iran, Turkey, the Mediterranean and a few oceanic ridges. Thus the earthquakes affecting the Himalayan region are a part of the global Alpide belt.

Effects of Earthquakes

An earthquake occurs whenever the earth shakes, irrespective of the cause. The effects produced are therefore a direct consequence of the motion of the ground. The effects are manifold and embrace man-made structures as well as the earth's surface. Within the epicentral tract, earthquakes of great intensity are known to cause considerable damage to buildings, railway lines, roads, bridges, canals, pipelines, rivers, fields and hillsides. Of particular significance are the effects in hilly regions. For instance, in the rainy season the mantle of soil and loose rock on hillsides is often sufficiently wet to be unstable. A slump occurs when a portion of the wet regolith on the hillside slips down a little, rotating somewhat on a horizontal axis parallel to the hillside.

Slumps frequently occur without earthquakes, but a large earthquake accelerates their formation and in a wet season their formation is a typical phenomenon. Huge masses of rock and debris thus thrust into river valleys form artificial barriers to river courses. This creates temporary lakes which eventually overflow and flood the countryside. A similar effect is also produced in rivers by earth or mud flows caused by earthquakes. The severe shaking that underground water-bearing strata are subjected to and consequent displacements within them often cause water to be forced up through open fissures. Water so expelled may render the soil fluid, so that it flows as a liquid. Such flows occur at times of shock and do not recur. Large earthquakes also bring about changes in the general topography of the epicentral region, including the levels of river beds. These could result in changes in the course of rivers.

Severe earthquake shaking of alluvial bottom lands frequently causes a lurching of the earth, accompanied by the formation of cracks, usually parallel to the stream. The Himalayan region comprising hills, valleys, and earthquake sources, is thus open to a variety of earthquake effects. Many of these effects have in fact been observed as may be seen from the following short notes on a few important events.

Kashmir Earthquake, 30 May 1885
(Epicentre: a few kilometres west of Srinagar)

This earthquake struck Srinagar within a few minutes of 2.45 a.m. at which time a pendulum clock, set to local time, stopped. The shock was severe and was felt in the Kashmir valley where much loss of life and damage to property occurred. The shock was also felt to a lesser extent in all the surrounding country and at Simla, Lahore, Peshawar, etc. Many aftershocks of less intensity followed the main shock, and were reported as late as in the latter half of July 1885.

According to a report by E.J. Jones, A.R.S.M., G.S.I., 'the number of persons killed by falling buildings according to official reports was something over 3000, while the number of cattle, horses, etc. killed was very great.' The shock was severe enough to do serious damage to buildings extending from the neighbourhood of Srinagar in the south-east, to a little north of Sopur and from near Baramulla down the Jhelum valley as far as the fort of Chikar near Garhi. The area over which extensive damage occurred was estimated to be about 750–1000 sq. km.

At Baramulla and higher up the river, a number of earth fissures were formed along the river banks and occasionally at some distance from the river. At Patan, the fissures varied from 2.5 cm to 90 cm in width. Jones reports he was informed that there were some much larger fissures a short distance from Baramulla, some of them being '20 yards in width and a quarter of a mile in length'.

A large landslip occurred at Larridar, about 11 km south of Baramulla. In the neighbourhood of the fissures there were numerous patches of fine sand which were forced up from some distance below the surface. They varied from 60 cm to 150 cm in diameter. According to the villagers, the sand gave off a sulphurous smell.

From various observations on the field, the position of the seismic focus was determined by Jones to be about 19.50 km west of Srinagar.

Assam Earthquake, 12 June 1897
(Epicentre: 26°N, 91°E)

This is one of the greatest earthquakes on record to date, and occupies a unique place in the development of seismology, not merely because of its force and the destruction it caused, but also because of the thorough and masterly use made of every bit of available field evidence in the investigation of this shock by R. D. Oldham of the Geological Survey of India. In Oldham's words:

At about quarter past five in the afternoon of the 12 June 1897, there burst on the western portion of Assam an earthquake which in violence and extent, has not been surpassed by any of which we have historic record. Lasting about two and a half minutes, it had not ceased at Shillong before an area of 150,000 square miles had been laid in ruins, all means of communication interrupted, the hills rent and cast down in landslips, and the plains fissured and riddled with vents, from which sand and water poured out in most astounding quantities; and ten minutes had not elapsed from the time Shillong was laid in ruins before about one and three-quarter millions of square miles had felt a shock which was everywhere recognized as one quite out of the common.

For an earthquake of its magnitude, however, the loss of human life was comparatively small, about 1500, because most of the inhabitants of the epicentral region were out of doors at the time.

The most violent part of the shock lasted probably a little less than one minute, with a total duration of perhaps three minutes. People were thrown to the ground by the shock; some were injured. The intensity of the shock was so great that waves were visible at a number of places, viz. Shillong, Nalbari and Mangaldai. Estimates of the horizontal ground acceleration at various places made from observations on pillars and monuments showed very large values, such as at

Silchar	:	1200	cm/sec^2	(1/8 gravity)
Dhubri	:	2700	,,	(2/7 ,,)
Cherrapunji	:	3000	,,	(3/10 ,,)
Shillong				
Sylhet }	:	4200	,,	(2/5 ,,)
Goalpara				

The vertical acceleration was so great that boulders were thrown upward, leaving cavities in the earth in which they had lain with sides almost unbroken. Such movement was observed at points 160 km apart. Many houses sank into the soft soil till only their roofs were visible.

The earthquake caused visible movements along faults, besides fracturing long stretches of rock. The Chedrang fault, located about 160 km west-north-west of Shillong, and running nearly north-south broke for about 19.30 km. The maximum displacement was 11 m vertical, the east side being elevated in relation to the west.

Landslides were very common and disastrous in the epicentral region; on the southern slopes of the Garo and Khasi Hills they were especially prevalent. Hills were stripped bare of their forest cover by slides for some thirty kilometres. Fissures in the alluvium were common over an area of about 650 km by 550 km. Earthquake fountains ejected sand to such as extent as to subsequently hinder farmers in cultivation.

The earthquake was followed by a very large number of aftershocks which were scattered over a large area. Up to the end of 1898, more than 5000 aftershocks occurred. Aftershock activity continued for about ten years after the main shock.

Kangra Earthquake, 4 April 1905

(Epicentre: 32.25°N, 76.25°E)

Closely following the Assam earthquake of 1897, the Kangra earthquake ranks as one of the most devastating to affect India. It had its origin in the Kangra district of the north-west Himalaya. According to Middlemiss,

Beginning at an early hour of the morning, when many people were still asleep, the more violent phases of the shock dealt summary destruction to life and property in the neighbourhood of the Kangra valley and Dharamsala; accomplished very great damage and caused considerable loss of life in the hilly tracts of Mandi state and Kulu; did serious damage to Dehra Dun, Mussoorie, Chakrata and other towns in the vicinity and slight damage to the large towns of Lahore, Amritsar, Jullunder, Saharanpur, and others similarly placed with reference to the centre.

The loss of life was 20,000 and the shock was felt over an area of 4,210,000 sq. km. Not only seismic instruments at Calcutta, Bombay and Kodaikanal in India but those in the Far East, Europe and America recorded the event, an indication of how powerful a shock it was. The meizoseismal area including Kangra, was on the tertiary rocks of the foot-hills of the Himalaya. The shock was so severe in the Kangra valley and the Dharamshala neighbourhood that people were thrown to the ground, buildings collapsed almost instantaneously and mortality was very high. At Kulu, people had to cling to trees for support, or were sent sprawling.

Although the cause of the earthquake was ascribed to movement of rocks below the surface of the ground, such movement nowhere occurred on the actual surface. There were however considerable secondary effects of the shock in the shape of fissures and landslides. In the immediate vicinity of Kangra and Dharamshala, fissures were caused along many of the slopes. In special cases such as at McLeodganj bazaar and at 'Byrn', marked destruction, numberless fissures, and actual subsidence of the land occurred. At Palampur one very noticeable rock-slide in the bare steep crystalline axis of the Dhauladhar range continued active for months after the quake. At stream-outlets along the Dhauladhar, the shedding of scraps of gravel terraces, and the skinning of steep slopes, carried away with them many miles of water-channels used for irrigating the lower parts of the Kangra valley.

At Larji, near which the confluence of the Beas with the Tirthan and Sainj streams takes place in deep gorges, the havoc wrought along the steeply-convex spurs was appalling. Both the Tirthan and the Sainj streams were temporarily dammed by debris cones, forming lakes. The water supply of Jawala Mukhi which has its source in springs was almost doubled. The boiling springs at Manikaran were affected slightly. One was blocked up altogether and left a public bathing place dry, whilst others within a short range shifted their channels. The springs at Mackinnons' Brewery at Mussoorie increased their discharge by 20 to 30 per cent and showed a gradual falling off only after 20 May.

The Dhubri Earthquake, 3 July 1930
(Epicentre 25.8°N, 90.2°E)

This is one of the earthquakes which seriously affected the northern districts of Bengal and adjoining areas in Assam. It struck during the early hours of the morning of 3 July 1930. Centering near the north-western end of the Garo Hills and the adjoining valley of the Brahmaputra river, the earthquake had disastrous effects in northern Bengal and western Assam, and was felt very distinctly over a wide area, extending from Dibrugarh and Manipur in the east, to Chittagong and Calcutta in the south, Patna in the west, and beyond the frontiers of Nepal, Sikkim and Bhutan in the north. In the epicentral tract which included the town of Dhubri, the shock attained an intensity severe enough to damage or destroy almost all the masonry and brick buildings. An interesting aspect of the damage to taller houses was that wherever the supporting posts were carried down into the plinth or into the ground, the structures were badly damaged—large cracks occurred in the plinths and the general vibration of the walls was sufficiently intense to cause heavy almirahs standing nearby to fall over. On the other hand, the few buildings in which the wooden framework merely rested on the plinth were undamaged, apart from slight cracks in the plaster. All buildings located on made-up ground, in filled tanks and ditches, etc. were severely affected. In several instances, where the structure had its foundations partly on the natural alluvium and partly on old excavations that had been refilled with earth, a large crack occurred along the junction of the in-filling and the alluvium, and the portion of the building located on the filling subsided over a metre.

At the time of the main shock, numerous fissures were formed in the alluvium ground, and sand and water spouted up from them to a height of several metres. Wells in the locality overflowed and were silted up by sandy alluvium brought up from below.

No loss of life was, however, reported.

The Bihar–Nepal Earthquake: 15 January 1934
(Epicentre: 26.50°N, 86.50 E)

Next to the earthquake in 1897 in Assam and the Dhubri earthquake in 1930, this is the earthquake of greatest intensity to have been experienced in Bihar and Nepal. The area of greatest devastation was in north Bihar and Nepal, but damage of gradually-diminishing intensity extended into the adjacent states. The epicentre of the shock lay below the alluvium slightly to the north of the towns of Darbhanga and Muzaffarpur. There were three zones up to 60 km in length which experienced the earthquake, the first west-south-west east-south-east from Motihari to Sitamarhi; the second at Monghyr south of the Ganga and the third around Kathmandu in Nepal. Within three minutes Monghyr and Bhatgaon were in ruins as also were large parts of Motihari, Muzaffarpur, Darbhanga and Kathmandu. In Sitamarhi, Madhubani, and Purnea, houses tilted and sank into the ground. One belt of ground which subsided extended from Bettiah in the north-west to Purnea in the south-east, a distance of nearly 320 km surrounding the epicentral tract. Subsidence of land was very widespread and throughout the slump belt there were innumerable fissures through which large quantities of sand and water were thrown to the surface, thus ruining the standing crop and making the soil unfit for cultivation. There were extensive landslides in the Himalaya towards the north.

The earthquake took a toll of more than 10,000 lives of which more than 3000 were in Nepal. It was felt up to a distance of 1600 km.

The Assam Earthquake, 15 August 1950
(Epicentre : 28.5°N, 97.0°E)

Second only to the Assam earthquake of 1897, this is the most intense earthquake to have occurred in recent times in India. Although the epicentre of the shock was near Rima, in a region close to the borders of India, China and Burma, this shock caused more damage in Assam, in terms of property, than the earthquake of 1897. There were severe floods, the rivers bringing down sand, mud, trees and all kinds of debris. Pilots flying over the meizo-seismal area reported great changes in topography, largely due to the enor-mous slides. Immediately after the primary shock several tributaries of the Brahmaputra river, particularly the Subansiri, Dihang and Tiding were blocked by landslips caused by the violence of the earthquake. The slip-dam of the Subansiri resisted for four days and ultimately burst, causing wide-spread havoc to the east of the Subansiri, inundating about 800 sq. km and thus causing great destruction of villages. This sudden flood took a toll of about five hundred lives. Considerable changes in the navigable channels of the Brahmaputra took place, particularly near the hill ranges. Fissures and sand vents were extensive due to the alluvial nature of the country.

An area of nearly 46,000 sq. km in Assam suffered from extensive and heavy damage. The shock was felt up to Lucknow, Allahabad and Rangoon. The total area affected must have exceeded 2,927,000 sq. km.

The Kinnaur Earthquake, 19 January 1975
(Epicentre: 32.5°N, 78.4°E)

The earthquake struck parts of the Kinnaur and Lahul-Spiti districts in Himachal Pradesh at about 1.32 p.m. on 19 January 1975. The shock caused severe damage near its source which was close to the Tibet border. In Kaurik, one of the worst-affected places, forty-two persons were reported killed and another forty injured. It destroyed 278 houses and damaged another 2000.

The quake triggered off heavy landslides upstream of Maling in the Spiti Valley. A landslide with a volume of 3.6 million cubic metres formed a dam across the Para Chu river about one kilometre upstream of Sumdo. Ground fissures appeared in the area closest to the epicentre, were 45 to 60 cm wide, and showed differential movement. Examination of the damage done to houses showed that those built of stone with wooden-frame structures, single-storey houses, and those constructed on rocky foundations suffered little damage.

This earthquake is second in severity only to the Kangra earthquake of 1905 in Himachal Pradesh, and is associated with movements along the Kaurik fault.

MECHANISM AND CAUSES OF EARTHQUAKES

We have seen how large a variety of effects are produced by earthquake motions. They underline the necessity for measures to safeguard life and property from such hazards. While this could be attempted in reconstruction work at those places which have been affected, there is also need to anticipate them, at least in certain regions. This requires a better understanding of the causes of earthquakes, and earth-scientists, and seismologists in particular, have been studying this problem for some time. Oddly enough, the first break-through was caused by the effects of an earthquake itself.

Reference has been made earlier to faulting on the ground with relative displacements on either side of the break. In the 1906 California earthquake, such a movement was clearly observed in the region of the San Andreas fault. Over a period of years before the 1906 earthquake, the US Coast and Geodetic Survey had made triangulation surveys of the area and they were repeated after the earthquake. The comparison of the pre-earthquake results with post-earthquake results showed that monuments over 48.3 km from the San Andreas fault had not moved during the earthquake. It was found that displacements were maximal at the fault and decreased with distance, so that a previously straight line became curved. With the help of these observations, Reid explained the immediate cause of the earthquake as an elastic rebound. His elastic rebound theory is simply illustrated in Figure 7.3.

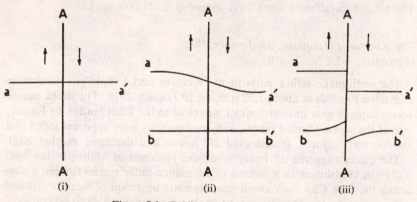

Figure 7.3   Reid's elastic rebound theory

In Figure 7.3 (i) aa′ is a line crossing the surface trace of the fault AA long before an earthquake. As inferred from the actual movements in earthquakes the differential forces acting on the two sides of the fault AA are shown by arrows. Figure 7.3 (ii) shows the state after a lapse of time, when under the stress of forces exerted on them blocks have been displaced, but at the fault are locked to each other by forces of friction. The original line aa' has been deformed. The other line bb' shows a fresh line introduced at this stage. The third stage (iii) shows the picture that emerges after the displacements have increased and the frictional or other forces holding the two sides locked across AA have failed. The line of markers bb' introduced in the intermediate stage has also broken into two parts with an offset at the fault, but also shows curvature. This last curvature fits the results of the re-survey after the 1906 earthquake (Figure 7.4). Thus it is generally accepted that the release of elastic strain is the ultimate cause of most earthquakes. Whenever such a state

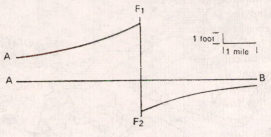

Figure 7.4

comes into being, the release will naturally take place at the weakest point. These are the faults which already exist and are mapped by geological and geophysical methods. The occurrence of earthquakes can therefore be related to faults and the forces which accumulate strains, to be released ultimately through earthquakes.

We shall consider the source of the forces later but the past earthquake history of the Himalaya does not leave any room for doubt that such forces do operate there. The results of surveys by the Geological Survey of India and others also reveal numerous fractures and faults across which the crust in the region has undergone movements. There is therefore enough basis to foresee the occurrence of earthquakes in these areas in the near future.

EARTHQUAKE ZONES

The epicentre map in Figure 7.2, although conveying a detailed picture of the Himalayan region, is not uniformly detailed for the entire area stretching from Kashmir to Assam. A close examination reveals that in the past two hundred years, earthquake activity has been more severe in the Hindu Kush and its vicinity, affecting Kashmir, the area around the Kangra valley, the trijunction of India, Nepal and Tibet, the east Nepal-Bihar border and the whole of Assam including the Shillong plateau. Although faults exist all along the Himalaya, they are not equally active. Taking into account earlier seismic activity and giving due consideration to known faults and their movements, it has been possible to prepare a seismic zoning map of the region. Such a map, prepared under the auspices of the Indian Standards Institution, is shown in Figure 7.5. The map is self explanatory; the areas where the seismic

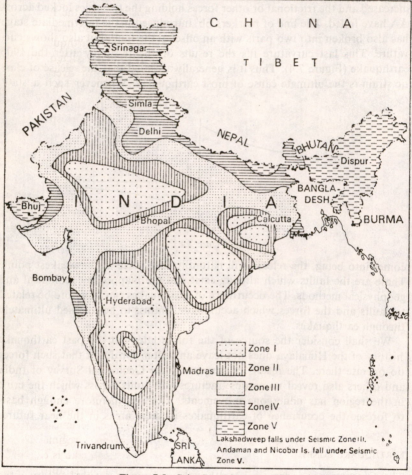

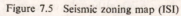

Figure 7.5   Seismic zoning map (ISI)

hazard is assessed at the highest are shown in Zone V, with a comparatively lower rating given to others. The intensity ratings of these zones are:

Zone	Intensity
V	IX and above
IV	VIII
III	VII
II	VI
I	V

The specifications of the Intensity Scale are given in the Appendix. Suggestions regarding methods to be adopted in reducing loss to structures in future earthquakes, according to the zoning map, are given in the ISI publication No. 1893 (1975) entitled *Criteria for Earthquake-Resistant Design of Structures*.

A quantitative assessment of the seismic risk, in terms of the frequency of occurrence of earthquakes of different magnitudes has also been attempted. Such assessments depend on statistical analysis of past earthquake data. Analysis of the instrumental data available since 1916 in the Meteorological Department shows that an earthquake of magnitude eight (or over)—which is equal in severity to either the Kangra, the Bihar–Nepal or the Assam earthquakes—on the average is expected to occur once in 38 years. The recurrence intervals of earthquakes of lesser magnitudes are correspondingly smaller, as shown in the table below.

Area	Expected time (years) for recurrence of earthquakes of magnitude				
	4	5	6	7	8
Himalaya	—	—	0.44	3.6	38
250 km around Dehra Dun	0.07	0.46	4	—	—
250 km around Chatra	0.11	0.71	4.3	—	—

The table also gives the results for a 250 km-radius circular area around Dehra Dun and Chatra (in Nepal) for which seismological data is available. The information given by these analyses, has, however, to be taken only as long-period average expectancies. In any specific period the events may succeed one another either more or less frequently.

It may be noticed that the table does not include the expectancies for larger magnitude events for the whole of the Himalaya. This does not imply that such events are not expected. In fact, according to the law of earthquake occurrence used in these analyses, smaller earthquakes are more frequent than larger ones. Therefore wherever a magnitude 6 earthquake takes place, those of magnitude less than 6 also occur. But a complete monitoring of the minor events is not possible without a dense network of seismic detection stations, and without uniform data, extrapolation of such regimes is not justified. Similarly in any region where a magnitude 6 earthquake is expected, a larger earthquake will have a lower expectancy. But such extrapolation to larger earthquakes without actual data from the region is also not possible, for the strongest earthquake which can occur on any fault system depends on many factors. This is a problem which has not yet been fully solved.

RECENT DEVELOPMENTS

We have seen earlier that earthquakes are associated with the breaking of rock masses across faults when the forces exceed the limits which the fault

can withstand. Where do these forces come from? An understanding of their
source, being basic to the understanding of earthquake occurrence, has
been attempted from very early times. Of the various theories in this connection the following may be mentioned:

Contraction of the Earth. Most theories of the origin of the earth assume that
it has cooled from a molten mass. Now it is known from the study of earthquake waves penetrating the earth's body that it is solid at least up to a
depth of 2900 km. Cooling, though a slow process, must have been greater
at the outer surface than in the deeper layers which are warmer. The upper
crystal layers therefore receive some heat from the deeper layers as well as
from the sun. Since the surface layer is in equilibrium and does not show any
change in temperature, it does not change in size, and tends to become too
large to fit the cooling and shrinking layers below. It must collapse to
shorten.

Isostasy. About 200 years ago it was observed that the mountain mass of
Chimborazo in Peru did not attract a suspended pendulum by as great a
force as it should have if the mountain were an added mass on the earth's
surface. Similar results have also been observed in various parts of the world
and they show that most mountains have an excess of lighter material below
them than the adjoining regions. In other words, like an iceberg floating on
water, with a part of its lighter mass below the water-line, mountains also
have roots. The theory of isostasy which evolved from these observations
states that any excess of matter on the earth's surface above a standard level
is balanced by a defect below it and vice versa. This defect can be brought
about by a flow of material away from below the mountains. This tendency
for the maintenance of gravitational balance in the earth's crust would offer
an origin for that slow accumulation of strains required by the elastic rebound
theory.

Continental Drift. The principal exponent of shifting continents was Alfred
Wegener. Proceeding from the observation that if South America were
moved eastward to rest against Africa, the eastern coast line of the former
would fit fairly well with the western coast line of the latter, Wegener framed
his hypothesis that all the land masses of the earth were grouped together in
one great continent up to the Cretaceous period (150 million years ago).
Then rupture began and different parts started drifting away. Antarctica,
Australia and India adjoined Africa before they split away. India in its movement folded the continental mass between it and the present Central Asia
into the Himalayan highlands. Similarly on the forward side of drifting
continents great mountains are pushed up by such forces. If such a behaviour
is granted, we have a source for the slow accumulation of strain in many
earthquake regions of the world, although the theory suffers from an in-
ability to explain how the forces necessary for the drift arose.

Convection Currents. Convection currents inside the earth are used to explain the forces which cause an accumulation of strain in the crust. The

elliptical shape of the earth produces thicker outer shells near the equator and therefore greater radioactive heating. Excessive heating and blanketing near the centres of continents causes convection currents towards the coasts with a resultant tendency to rupture continents. It was also envisaged that in some way these convection currents were met by opposing currents coming from the ocean at the continental boundary. The meeting of these two, which then plunge down, compresses the continental borders, throwing up mountains. The problem in this theory is the difficulty in explaining how the heating necessary to start the cycles arose, and explaining its cooling period as well.

Spread of the Ocean Floor and Plate Tectonics. During the last fifteen years much geophysical work has been done throughout the world under the International Upper Mantle Project at the conclusion of which another project, the Geodynamics Project, was taken up. The data collected included those from seismological, geomagnetic, heat flow, geochronological and other geophysical fields. The synthesis of these data has resulted in a theory called Plate Tectonics.

From a study of earthquake waves which penetrate the earth's interior, it is known that the earth has a crust about 35 km thick under continents but thinner under oceans. Below the crust is a thick shell called the mantle, down to a depth of 2900 km. The core inside is in a liquid state, but it is believed that it too has a solid inner kernel. The density of the material increases with depth as does the temperature. A cross section of the earth as is known at present is shown in Figure 7.6.

Recent research has further shown that although the mantle is solid, a thin layer at the top—roughly between 100 to 200 km from the surface—is in a comparatively plastic condition allowing slow movement over it. Earthquakes occurring in some of the island arc regions have also shown that they occur along a slightly dipping layer which goes under the continents. The motions of the crust computed from a study of earthquake-recordings confirm that they are either (1) thrusting two blocks one against the other, (2) rifting the blocks, or (3) sliding one past the other. Thrust forces have so far been observed in the island arcs and also in the Himalayan region. The mid-oceanic ridges show rifting whereas slipping forces are observed in other regions.

Studies of the magnetic properties of rocks on the ocean floor, as well as their age, have revealed that the ocean floor is spreading across ridges. These and similar observations from other branches of the earth sciences have been explained by a system of plates on the surface of the earth moving in relation to one another and maintaining the overall spherical surface of the earth. The plates which thus float on the plastic zone mentioned earlier are called lithosphere plates and the underlying layer in the mantle is called the asthenosphere. The nature of movements indicated is shown in Figure 7.7.

According to this theory of Plate Tectonics, the earth consists of a small

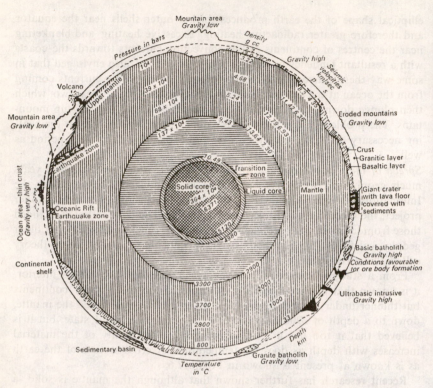

Figure 7.6   Diagram depicting the interior of the earth

number of plates. Their movement gives rise to forces at the plate boundaries which result in earthquakes. India lies on the Indian Plate which is moving in a northerly direction meeting the Eurasian Plate in the Himalayan Region. The evolution of this mountain range is ascribed to their collision. Earthquakes in this region also owe their origin to this collision.

Incidentally, this concept of the ocean floor spreading and moving plates is a revival of the formerly abandoned theory of Continental Drift, which is now explained by the results of the latest studies. In fact, working backwards in time from the present position of the continents, it has been confirmed that all the continents together formed one single land mass called Pangaea about 200 million years ago. Initial rifting about 180 million years ago separated the land mass into two: Gondwanaland, comprising South America, Africa, India, Australia, Antarctica; and Lanrasia with North America and Eurasia. Subsequent rifting separated India from Africa, Antarctica and Australia, and moved it northwards. As it drifted and approached the present Asian continent it narrowed the intervening sea, the Tethys. Finally it collided with the Asian continent, the Tethys sea gradually filled

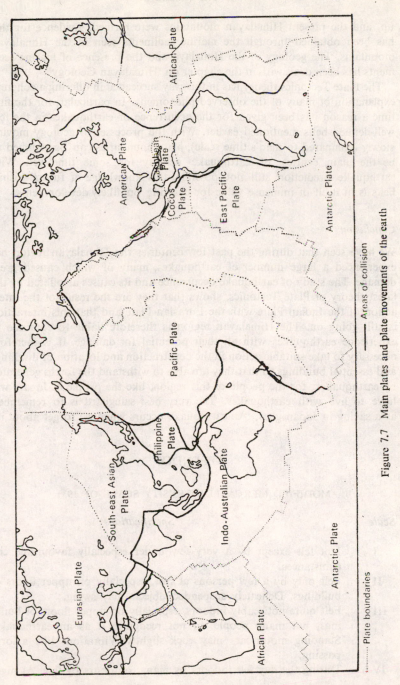

Figure 7.7 Main plates and plate movements of the earth

········· Plate boundaries

—— Areas of collision

up, and the present Himalayan mountains were formed. Evidence for this has been obtained through the marine sediments seen in the Himalayan mountains. The geology of the Himalaya and the presence of marine sediments has been dealt with in the chapter on 'Himalayan Geology'.

The Plate Tectonics theory has in this way succeeded in offering a coherent explanation of many of the observed phenomena. In particular, for the first time a reason has been given for the occurrence of earthquakes in the few well-defined belts mentioned earlier. With the processes of geology moving slowly (compared to man's time scale), plate boundaries can be assumed to be the main centres of earthquake activity for some time yet. With earthquake prediction still not possible, this knowledge of future seismic belts is in itself of immense value for charting out likely developments.

Conclusion

We have seen that during the past few centuries the Himalayan region has experienced a large number of earthquakes, many of which caused great damage. The study of earthquake occurrence and its causes as offered by the latest theory of Plate Tectonics, shows that they are the result of the interaction of the Indian Plate with the Eurasian Plate and that this interaction is still going on. The Himalayan region is therefore liable to continue to experience earthquakes with all their potential for damage. It is therefore necessary to take suitable action in the construction and location of dwellings and essential buildings so that they are able to withstand the forces generated by earthquakes; for the people of this region, like the people in Japan, will have to live with earthquakes. The very best safeguard is to remember, as a saying in Japan goes, 'An earthquake occurs when we forget about it.'

APPENDIX

MODIFIED MERCALLI INTENSITY SCALE OF 1931

Scale	Specifications
I	Not felt except by a very few under especially favourable circumstances.
II	Felt only by a few persons at rest, especially on upper floors of buildings. Delicately suspended objects may swing.
III	Felt quite noticeably indoors, especially on upper floors of buildings, but many people do not recognize it as an earthquake. Standing motor-cars may rock slightly. Vibrations like a lorry passing.
IV	During the day felt indoors by many, outdoors by few. At night

some awakened. Dishes, windows, doors disturbed, walls make cracking sound. Sensation like heavy lorry striking building. Standing motor-cars rocked noticeably.

V   Felt by nearly everyone; many awakened. Some dishes, windows, etc. broken; a few instances of cracked plaster; unstable objects overturned. Disturbance of trees, poles and other tall objects sometimes noticed. Pendulum clocks may stop.

VI   Felt by all; many frightened and run outdoors. Some heavy furniture moved; a few instances of fallen plaster or damaged chimneys. Damage slight.

VII   Everybody runs outdoors. Damage negligible in buildings of good design and construction; slight to moderate in well-built ordinary structures; considerable in poorly built or badly designed structures; some chimneys broken. Noticed by persons driving motor-cars.

VIII   Damage slight in specially designed structures; considerable in ordinary substantial buildings with partial collapse; great in poorly-built structures. Panel walls thrown out of frame structures. Fall of chimneys, factory stacks, columns, monuments, walls. Heavy furniture overturned. Sand and mud ejected in small amounts. Changes in well water. Disturbs persons driving motor-cars.

IX   Damage considerable in specially designed structures; well-designed frame structures thrown out of plumb; great in substantial buildings, with partial collapse. Buildings shifted off foundations. Ground cracked conspicuously. Underground pipes broken.

X   Some well-built wooden structures destroyed; masonry and frame structures and their foundations destroyed; ground badly cracked. Rails bent. Considerable landslides from river banks and steep slopes. Shifted sand and mud. Water splashed over banks of rivers, etc.

XI   Few, if any masonry structures remain standing. Bridges destroyed. Broad fissures in ground. Underground pipe lines completely out of service. Earth slumps and landslips in soft ground. Rails bent greatly.

XII   Damage total. Waves seen on ground surface. Lines of sight and level distorted. Objects thrown up in the air.

BIBLIOGRAPHY

BANERJI, S. K. 1957. 'Earthquakes in the Himalayan Region'. Indian Assoc. Cult. of Science. Calcutta.

CHAUDHURY, H. M. 1973. 'Earthquake Occurrence in the Himalayan Region and the New Tectonics'. Paper read at seminar on: Geodynamics of Himalayan Region. National Geophysical Research Institute, Hyderabad.

——. H. N. SRIVASTAVA and J. V. SUBBA RAO. 1975. 'Seismotectonic Investigations of the Himalaya'. *Himalayan Geology* vol. 4.

COX, ALLAN. 1973. *Plate Tectonics and Geomagnetic Reversals*. W. H. Freeman. California.

GANSER, A. 1964. *Geology of the Himalaya*. Interscience Publishers. London.

GUTENBERG, B. and C. F. RICHTER. 1954. *Seismicity of the Earth and Associated Phenomena*. Princeton University Press.

Indian Standards Institution. 1976. *Criteria for the Earthquake Resistant Design of Structures*. Delhi. (IS 1893).

KRISHNAN, M. S. 1960. *Geology of India and Burma*. Higginbothams, Madras.

Memoirs of the Geological Survey of India. 1883. Vol. 19. Pt. 3.

——. 1899. Vol. 29.

——. 1901. Vol. 30.

——. 1910. Vol. 38.

RASTOGI, B. K., J. SINGH and R. K. VERMA. 1973. 'Earthquake Mechanism and Tectonics in Assam-Burma Region'. *Tectonophysics, 18*, 1973.

RASTOGI, B. K. 1974. 'Earthquake Mechanisms and Plate Tectonics in the Himalayan Region'. *Tectonophysics, 21*, 1974.

Records of the Geological Survey of India. 1885. Vol. 18, Pt. 4.

——. 1905. Vol. 32, Pt. 3.

——. 1905. Vol. 32, Pt. 4.

RICHTER, C. F. 1958. *Elementary Seismology*. W. H. Freeman. California.

SRIVASTAVA, L. S., P. SINHA and V. N. SINGH. 1974. 'Tectogenesis and Seismotectonics of the Himalaya'. 5th Symposium on Earthquake Engineering, India.

TANDON, A. N. and H. N. SRIVASTAVA. 1975. 'Focal Mechanisms of Recent Himalayan Earthquakes and Regional Plate Tectonics'. *Bull. Seismological Soc. America, 65*, 1975.

WADIA, D. N. 1957. *Geology of India*. Macmillan. London.

Wadia Institute of Himalayan Geology. *Himalayan Geology* vols. 1–5.

TANDON, A.N. and S.N. Chatterjee, ·Seismicity Studies in India, *Indian Journal of Meteorology and Geophysics*. Vol. 19, No. 3, 1968.

CHAPTER 5

SOCIAL CHANGE IN A HIMALAYAN REGION

CHRISTOPH VON FÜRER-HAIMENDORF

The Himalayan regions comprised today within the Kingdom of Nepal for centuries played the role of a contact zone where Hindu civilization and the way of life of Tibetan Buddhism met. Though throughout the hundred years of the Rana era Nepal tended to isolate itself from foreign and particularly western influences, its people maintained a variety of commercial links with both Tibet and India, and some of the principal trade routes between its two great neighbours traversed Nepalese territory, following the courses of rivers flowing from north to south, and crossing the Himalayan main range by passes of great altitude.

One of these trade routes ran through the valley of the Kali Gandaki, and from there through the territory of the Raja of Mustang, known to his own subjects and Tibetans in general as *Lo Gyelbu,* the King of Lo. Several different peoples used to be involved in the flow of goods along this route, and the decline of the trans-Himalayan trade caused partly by the Chinese take-over of Tibet, and partly by developments inside Nepal, has led to profound economic and social changes throughout the Kali Gandaki zone. On the basis of anthropological investigations which I began in 1962 and continued in 1976, I propose to analyse these changes and their effect on the various populations traditionally dependent on the trade between Tibet, Nepal and India.

The part of the Kali Gandaki valley hemmed in by Annapurna and Dhaulagiri is known as Thak Khola, and its inhabitants are generally described as Thakalis, though in their own Tibeto-Burman language they call themselves 'Tamang', without however implying any connection with the Tamang tribe of Eastern and Central Nepal. The Thakalis are fundamentally a trading community, and for long periods they enjoyed a virtual monopoly of the salt trade following the Kali Gandaki route. Concentrated in

I gratefully acknowledge the financial support of the British Academy and the School of Oriental and African Studies which enabled me to return to the scene of my field work of 1962, and assess the changes which have occurred during the fourteen years of my absence.

thirteen villages situated on both sides of the river at altitudes between
2135 m and 2440 m, the Thakalis operated a mixed economy based on trade,
agriculture and some animal husbandry, including the breeding of yak. The
main trading centre was Tukche, a settlement of close on a hundred houses
built on a level site close to the banks of the Kali Gandaki. It served as an
entrepôt where wealthy Thakali traders stored in enormous palatial houses

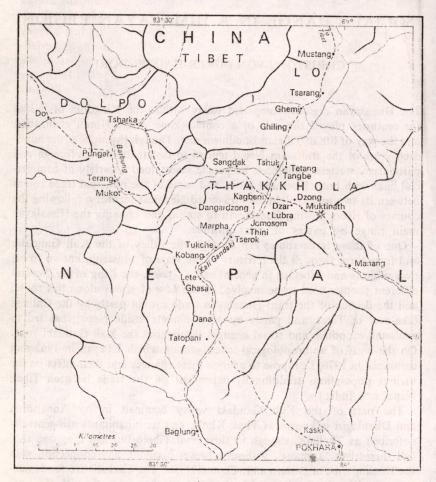

the salt and wool imported from Tibet, as well as large quantities of grain
grown in the middle ranges of Nepal, and carried to Tukche by the growers
who bartered it for salt at a rate extremely favourable to the Thakali merchants. The latter subsequently sold the grain in Tibet, usually in exchange
for salt and wool, and derived large profits from this chain of transactions.
  When I first studied the Thakalis the prominent merchant families of Tukche

still dominated the whole region—not only the thirteen Thakali villages of the region known as Thaksatsae (Thak of the seven hundred [houses])— but also the areas lying to the north known respectively as Panchgaon and Baragaon. A council of thirteen village headmen claimed control over the social affairs of Thaksatsae, but within this council the principal families of Tukche exercised virtually unlimited authority. Yet, the power of the Thakali merchants had even then passed its zenith, for the trans-Himalayan trade was beginning to suffer from the impact of Chinese rule over Tibet. Moreover, several of the more prominent Thakali merchants had already extended their operations to areas outside Thak Khola, and some had established themselves in Kathmandu, Pokhara and certain localities in the *tarai*. While previously Thakalis used to spend only some of the winter months in regions of lower altitudes, those with commercial establishments in such towns as Pokhara and the Bhairawa gravitated more and more towards the middle ranges and lowlands and returned only for limited periods to Thak Khola. Yet, the social fabric of the Thakali community of Tukche and the neighbouring villages was still unimpaired, and the leading families continued to exert their influence on the regions lying to the north and north-west of Thaksatsae.

When I returned to Thak Khola in 1976, fourteen years after my first stay among the Thakalis, the situation had changed completely. The majority of the Thakalis of Tukche had moved to Pokhara and other places in the middle ranges, and Tukche had the appearance of a dying town. Many houses were locked up, in others there were only caretakers, and there was a whole quarter where most of the houses, deserted by their owners, had totally collapsed. Whereas in 1962 the population was 495 it had by 1976 shrunk to 223, and even this figure was deceptive because many so-called residents spent only brief spells in Tukche, and had their main businesses and houses in Pokhara or in one of the towns of the *tarai*. In Pokhara alone there were about 85 Thakali households, representing a population of about 425, while at the time of my visit (February–April 1976) there were only 33 true Thakalis staying in Tukche. The population consists now mainly of non-Thakalis, the offspring of bond servants and dependents of the great Thakali families. Their parents or grand-parents stemmed either from Marpha and other Panchgaon villages, or from one or the other of the settlements of Baragaon or more rarely Mustang.

Some of them have acquired some land of their own, but most cultivate land still belonging to Thakalis, and pay a share of the crop in lieu of rent. As cultivators they appear to be rather inefficient, for having been used to the strict supervision of their former masters, they lack initiative. There is also a noticeable shortage of animal manure, for with the decline of trade the number of pack animals has diminished, and Tukche does not benefit any longer from the manure left behind by the huge caravans of sheep, goats, donkeys and *dzoppa* (a cross-breed between yak and ordinary cattle)

which used to bring trade goods to Tukche or halted there in the courtyards
and stables of the great houses of Thakali merchants. Today caravans travelling south from Jomosom, Baragaon or Mustang rarely stop in Tukche,
and trade is reduced to some local transactions in goods brought from Pokhara or Baglung.

It could reasonably be assumed that the present inhabitants of Tukche,
the majority of whom had for generations been dominated and often exploited by their Thakali masters, would now exert themselves to stand on
their own feet. Though much of the land they cultivate still belongs to
Thakalis who have moved away from Tukche, there should be sufficient
scope for improving their economic status. The new system of village government, the *panchayati raj*, gives them every opportunity of rising above their
earlier underprivileged position, for eight out of eleven members of the local
council of Tukche are men of humble origin and known as *arangsi-karangsi*,
in contradistinction to the established property-holding families, which
are known as *kuria*. Yet so far their representatives have made little effort
to make their voice heard and to step into the shoes of the former upper
class. A *panchayat* member by the name of Rapke, whom I knew well from my
earlier stay in Tukche, exemplifies the position of the majority of the present
inhabitants of Tukche. He had originally come from Marpha and is hence
not a Thakali in the narrow sense, while his wife came from a village in
Baragaon and worked for eighteen years as a bond servant of one of the
wealthy Thakali families. Rapke himself was also in the service of a Thakali
and he still lives as a caretaker in the house of his former master, who has
settled in the *tarai* in the vicinity of Bhairawa. Though he seems reasonably
well off and his eldest son is a teacher in the local school, he complained to
me that fundamentally his own condition and that of his class has not changed. He still had no house of his own and he cultivated Thakali land on terms
no more favourable than earlier ones. The owners were holding on to their
land even if they let their houses collapse, and the sharecroppers had established no rights on the fields they cultivated. Just at the time of my visit a
land survey was in progress and the provisions of the new land reform regu-
lation favour the actual cultivators *vis-a-vis* the absentee landlords, giving
them rights over a portion of the land. Yet, neither Rapke nor other cultivators of his class took any steps to have their rights recorded. The *panchayat*
which could have made representations to the survey party failed to do so
and Rapke explained that there was no popular support for such a move.
People regarded as *arangsi-karangsi*, a term for landless residents which has
always had a slightly derogatory implication, are still intimidated by their
masters, and their membership of the *panchayat* has failed to give them self-
confidence. Though in a majority on that body, they invariably follow the
lead of the chairman and vice-chairman, who are two of the few Thakalis
who have remained in Tukche.

These Thakalis are trying to revitalize the economy by developing fruit and vegetable farming, an enterprise greatly assisted by an agricultural farm situated at Marpha, two hours' walk north of Tukche. They have planted apples, peaches, walnut trees, grapes and lemons, as well as a variety of vegetables previously unknown in Thak Khola. The quality of fruit and vegetables is good, but marketing will present difficulties as the nearest motorable road is six days march away, and air transport from Jomosom would add at least Rs 2.25 to the cost of every kilogram of fruit transported to Kathmandu, where Kashmir apples are sold for Rs 5 to 6 per kilogram. So far the yield of the young trees is too low to allow experimentation in marketing, but the fruit-growers hope that in the long run the development of tourism will provide at least a partial answer to their marketing problems.

The prospect of fruit growing has led to a rise in land prices, but so far it has not dispelled the atmosphere of lethargy and decay which prevails in Tukche, once the most progressive and today the least prosperous of the villages of Thaksatsae.

The decline in the population and the economy has also affected the religious life of Tukche. While in 1962 there were still patrons to commission Buddhist rituals and recitations in which lamas from several surrounding villages participated, today there are few such performances. The three *gompa* are still being maintained, but except for one old Tibetan there are no resident lamas, and the only religious practitioners are a few elderly nuns who take turns in looking after the *gompa* in their charge.

Different conditions prevail in the villages south of Tukche, such as Kobang, Sauru, Larjung. These villages have relatively more and better land than Tukche, and agriculture has always played a more important role in their economy. Though most people had engaged to some extent in trade there had been no great merchant families comparable to those of Tukche. Hence the decline of trade with Tibet has had a less damaging effect, and fewer villagers have settled permanently in Pokhara or Bhairawa. In 1976 Kobang had a total population of 92, and of these 74 were members of true Thakali clans and only 18 were non-Thakalis. In Sauru, the respective figures were 26 against 24, and only in Larjung, which had a population of 74 did 41 non-Thakalis out-number the 32 Thakali residents. The general feeling in these villages is that the standard of living has gone down somewhat, but to the outside observer this is much less noticeable than in Tukche. There is certainly no sign of a general abdication by the men of the old Thakali clans, and indeed the members of the Kobang *panchayat* which includes the villages of Kobang, Larjung, Nabrikot, Nakung, Burjmukot, Khanti and Sauru, are without exceptions true Thakalis. The non-Thakalis have no representation on the *panchayat* even though in Larjung they form the majority of the population.

Those men who in the past had made large profits from long-distance trade such as the purchase of wool in Tibet and its sale to Indian textile mills, are depressed by the decline in their income, and some are talking of moving to Pokhara or other commercial centres. They say, however, that at present they cannot get a reasonable price for their houses and land in Thak Khola, and hence cannot raise the capital for the establishment of a business elsewhere. Expansion of agriculture in Thak Khola is apparently not practicable because all level land is already under cultivation and there is no point in clearing forest on hill-slopes. However, several men own herds of yak, and it seems that yak-breeding has increased in importance. A man who owns 30–40 yak can earn about Rs 3000–4000 per year, partly from the sale of butter and partly from that of young animals to the people of Dolpo, a high altitude region west of Thak Khola. The Bhotias of Dolpo use the young yak for breeding and carrying loads, and give in exchange fully-grown yak which the Thakalis in turn sell to villagers of Baragaon for slaughter. The people of the old Thakali clans no longer eat yak meat, which is considered beef, and hence shunned by those emulating Hindu customs. It is said, however, that some of the *arangsi-karangsi* eat yak meat, and this is cited by Thakalis as a reason for not liking to eat in the houses of their co-villagers of lower status.

Even though in villages such as Tukche the sons of immigrants and former bond servants are now *panchayat* members, the two strata within the society of Thaksatsae tend to retain their respective identities. There is little intermarriage between the members of the four original Thakali clans (Sherchan, Tulachan, Gouchan and Bhattachan) and those stemming from Panchgaon or Baragaon, or descended from Magars serving in Thakali houses. It seems that neither of the two classes favours mixed marriages, though matches not arranged by the parents but resulting from love affairs and elopements do occasionally occur. There are also differences in the ritual practices of 'old' Thakalis and the newcomers. While both join in the performance of Buddhist rituals, only the Thakalis of the four clans attend the worship of their respective clan deities. This invariably involves the sacrifice of animals, which is objectionable to devout Buddhists and hence to those coming from the purely Buddhist communities of Baragaon.

One of the bizarre consequences of the incipient political emancipation and aspirations of the former dependents of the Thakalis is the way in which they use the term 'Thakali' as a surname. Whereas members of the four Thakali clans refer to themselves by the clan-name, e.g. Govindaman Sherchan or Hira Bahadur Tulachan, residents in Thakali villages who hail from Panchgaon, Baragaon or from a Magar community add the name Thakali to their first name, e.g. Lal Bahadur Thakali. The result is that if a man calls himself Thakali one knows that he is not a Thakali in the true sense, whereas a-man adding the clan name Sherchan, Gouchan Tulochan, or Bhattachan to his name is at once recognized as a genuine Thakali. It goes without saying that persons of the latter type will also describe themselves as Thakali if specially asked for their caste by somebody not familiar with Thakali clan names.

The problem of nomenclature is further complicated by the relatively recent tendency of residents of the Panchgaon villages, a group of settlements north of Thaksatsae, to describe themselves also as Thakalis when speaking to outsiders. In their own language the people of Marpha village used to call themselves Punel, those of Thini village Thinel, and so on, but their increasing contact with people unfamiliar with the villages and ethnic groups of Thak Khola, has led them to refer to themselves as Thakali, no doubt with the intention of cashing in on the true Thakali's reputation as successful and wealthy traders. They are much more justified in doing this than Baragaonlis or Magars resident in Thaksatsae when they assume the name 'Thakali', for the general social structure and cultural traditions of the people of Panchgaon are very similar to those of the Thakalis, and even their language is closely akin to that of Tukche. But the Thakalis tend to consider themselves superior to the people of Marpha and the other villages of Panchgaon, largely because in the past they had been able to dominate them economically and politically, and some men and women of Panchgaon used to serve in the houses of rich Thakali traders.

However, Marpha, a village of 120 houses, which lies only four miles north of Tukche, has stood up to the economic changes of recent years far better than most of the villages of Thaksatsae. Unlike the Thakali society, that of Marpha was never a stratified community, and although there were always differences in the affluence of the various families there were no class distinctions comparable to those of Tukche. With very few exceptions all villagers belong to one of the four traditional clans of Marpha, and marry only among themselves. Endogamous and closely structured, Marpha is a compact community very conscious of its own identity. All land within the village boundaries is owned by Marphalis, and no Marphali owns land in any other village of Thak Khola.

At present Marpha is included within a *panchayat* which comprises also the two neighbouring Panchgaon villages of Shyang and Tsairo. Moreover an informal assembly consisting of representatives of all Panchgaon villages, i.e. Thini and Chiwog in addition to Marpha Shyang and Tsairo, has recently been established on the initiative of the Marpha *panchayat* chairman, and this assembly which is a consultative and co-ordinating body meets whenever matters of common interest are to be discussed. The assembly, like the government-sponsored *panchayat,* is a new institution, for in the old days each of the Panchgaon villages was self-contained and autonomous, and no body existed comparable to the council of the thirteen *mukhyas* of Thaksatsae.

Notwithstanding the new *panchayat* organization, the traditional form of village government has been largely maintained. It is based on a system of village officials representing the four clans (*tsho*) into which the community is divided. Each of the *tsho* elects one *thumi,* who acts as a clan-headman for one year. One of the four *thumi* is chosen as chairman, now known by the Nepali term *mukhya*. His position is not attractive, for the *mukhya* is under the obligation to stay throughout the year in Marpha, while other *thumi* may go on trading journeys during winter when the majority of the villagers have left for the middle ranges or lowlands. The *thumi* and particularly the *mukhya* supervise the work of eight other village officials known as *tshowa,* who act as messengers and guardians of public order. Two of the *tshowa* are described as *tsilowa,* and one of these acts as village accountant while the other organizes the work of the remaining *tshowa.* The main function of the *tshowa* is to guard the fields against cattle stealers and to keep the irrigation system in good order. They take turns in the discharge of these duties, one *tshowa* working every eighth day. The *tshowa* are appointed in rotation from among the landowning inhabitants (*kuria*) of Marpha, and any *tshowa* who leaves the village when his term of duty comes must provide a substitute or pay a fine. Fines are also exacted from eligible villagers who evade appointment as *thumi* by absenting themselves from the village.

In the days when the Thakalis of Tukche exercised a measure of authority over Panchgaon and Baragaon, one member of the leading family of Tukche held the position of *chikyap* of Marpha. This is a Tibetan title standing for 'chief' or 'headman', and in this capacity he confirmed the election of *thumi,* but nowadays the posts of *thumi* are virtually filled in rotation. In place of the *chikyap,* whose post though never formally abolished has become obsolete, the elected *pradhan panch,* i.e. the chairman of the government *panchayat* which comprises Marpha, acts as the highest authority in the village and selects one of the *thumi* to act as *mukhya.* On various ceremonial occasions when the villages divide into two parties, such as during a ritual dance on the occasion of the Mang-diva (casting-out of the evil spirit), the *pradhan panch* acts as the leader of one party while the *mukhya* leads the other party.

All these village officials are subject to criticism by the village community as a whole. Once every three years respected men representing the four clans meet for three days, scrutinise the actions of the village officials and revise if necessary the rules for the administration of the village. At this meeting *thumi, tsilowa* and *tshowa* may not be present, but they have to produce the records relating to their administration of village affairs. In theory they are entitled to fine village officers who have been remiss in the discharge of their duties.

Disputes within the village are brought before the *pradhan panch* who selects by lottery several prominent men, usually *thumi* or *panchayat* members, to help him in passing judgement. Up to 1976 no case has occurred in which the parties did not accept the judgement of this court, or where the litigants appealed over the heads of the village authorities to a government court outside Thak Khola.

Marpha appears to be an unusually well-governed community with none of the rivalries and bitter quarrels which had divided Tukche even in 1962 when its economy still supported a relatively high standard of living.

Today Marpha compares favourably with the villages of Thaksatsae even in the economic sphere. Though it lies at the southern end of a very arid zone and agriculture is possible only with the help of irrigation, the land is much more productive and much more carefully cultivated than that of Tukche. Few people start work much before 8 a.m. in Tukche and many can be seen sitting on the steps of their houses most of the morning, while in Marpha groups of men and women go to work long before sunrise. Moreover the owners of the land themselves work in their fields, though during the harvest they employ labour recruited in Baragaon. Unlike many Thakalis, they do not consider manual labour demeaning, and the wives and daughters of wealthy and prominent men—attractive women who appear beautifully turned out at festivals—can be seen carrying manure or large bundles of buckwheat or maize straw, while the men do the ploughing and threshing, without fearing any loss of status. Though seasonal hired labour is employed for some of the agricultural work, hardly anyone keeps a domestic servant, and there are no men or women from Baragaon or Mustang who work permanently in Marpha houses as they did in the houses of wealthy Thakalis. Marpha people, whether rich or poor, work hard and for long hours whenever there is work to be done, and they throw themselves with great verve into the celebration of feasts.

Marpha, no doubt, took longer than Tukche to catch up with modern developments, and in 1962 this was expressed in the lower literacy figures. But since then the Marphalis have made great progress. In 1976 there were the following qualified persons among the people of Marpha: two with B.A. degrees, one with a B.Sc. degree, one with a B.Ed. degree, one qualified engineer, one captain in the Royal Nepal army, five overseers in government employ, sixteen with school-leaving certificates. One Marphali was studying medicine and one radiology in Japan, four were in boarding schools in Kathmandu and two in Pokhara. In the Marpha middle school, four teachers were paid by government and one from village funds.

Progress has been achieved not only in the educational field but also in the economic sphere. The main stimulus to economic innovation came from a government agricultural farm established in 1967 on the initiative of Pasang Sherpa, a trained horticulturist, who ever since its inception has been the life and soul of the enterprise. In the outward-looking and well-organized community of Marpha he found ready support for his ideas, and the villagers not only co-operated in the work of establishing the experimental farm, but also set out on their own to develop the growing of fruit and vegetables. The results have been most encouraging: the government farm and village produce large quantities of vegetables, and there are substantial plantations of apples, peaches, apricots, almonds, and walnuts. The main newly-introduced vegetables are cabbage, cauliflower, carrot, beans, tomato and onion. Some of these, as well as potatoes, are transported by mule to Pokhara, where they find a ready market, but like the Thakalis the Marphalis hope for even greater profits from a revival of the tourist traffic which was disrupted for two years in 1974, as a result of clashes between government forces and martial Khamba exiles from Tibet.

Marphalis have long been active in the transport business, and in the past they provided much of the carriage for Thakali merchants. In 1976 there were 350 mules, 60 horses and 35 *dzoppa* in Marpha. The village people also owned about 400 yak, kept on high pastures, and seven flocks of goats of some 70 animals each. This figure was not typical, however, because in the previous year an epidemic of foot-and-mouth disease had decimated the herds, causing losses estimated at Rs 100,000. As there is no spare pasture land in Panchgaon, there is no possibility of increasing the number of animals, and yak have to be taken to distant grazing grounds belonging to villages such as Sangda, which have to be paid grazing fees.

The composition of the livestock of Marpha in 1976 was very different from that which I recorded in 1962. Then there were only 32 mules, but 220 horses, indicating that the Marphalis have switched from horses to mules as their principal pack-animals. The number of *dzoppa* was then 89 and the drop to 35 in 1976 is probably due to the changeover from trade routes in high-altitude regions, such as Mustang and Dolpo, to routes in the lower country unsuitable for *dzoppa*. Conversely only 70 yak were owned by Marpha people in 1962 while the number in 1976 had risen to over 400, an indication of an increased emphasis on cattle breeding as opposed to trade and transport.

The mule and pony caravans of the Marphalis ply nowadays not only along the traditional Kali Gandaki and Pokhara routes. Road-construction projects and other development programmes have created a temporary need for animal transport in other parts of the middle ranges, and the enterprising Marphalis have obtained a large share of the market. The most adventurous men have not been content with providing mule transport, but have acquired and learnt to drive motor trucks which they use on the new roads from Pokhara to Kathmandu and Bhairawa.

The decline of trade with Tibet has affected Marpha much less than the Thakali villages and in particular Tukche, for the economy of Marpha had never been geared predominantly to trade. A small volume of trade in Tibetan salt and wool with Mustang remains, but the trade with Tsarka and other villages of Dolpo was disrupted by the Khamba troubles and has not yet revived, neither is there any trade with Nanang. On the other hand Marphalis have begun to import *dzoppa* and *dzomu* (cross-breeds between yak and cow). They buy them from the Sherpas of eastern Nepal, and then drive them from Solu to Trisuli and thence via Pokhara to Thak Khola.

Efficient agriculture provided the basis for the diversification of the economy. In 1976 I was told that the Marphalis had large stocks of grain and were then consuming the 1974 harvest, while the 1975 one was still untouched. Prices of land were high—Rs 11,000 to 12,000 was being asked for a field of approximately 1 *ropani* (0.13 acres) close to the village, and Rs 4000–6000 for a field of equal size at a greater distance from the village.

Marpha's healthy economy combined with the inhabitants' determination to maintain their traditional cultural life is also benefitting the Buddhist institutions in the village. A large *gompa* built on a rocky spur dominating the centre of the village is kept in good repair, and contains a very respectable collection of Tibetan scriptures. When the *gompa* was built in its present impressive shape on the site of an earlier smaller shrine, a reincarnate lama persuaded the villagers to create a fund in support of the *gompa* services, and this fund amounting now to from Rs 40,000 to 50,000 is administered by a layman selected by the lamas in charge of the *gompa*. Some 50 fields belong to the *gompa* fund, and the rent obtained from them yields approximately Rs 4000 per year. A further Rs 1000 to 1500 is annually collected for the performance of a festival *dokhyap*, involving rites and sacred dances extending over several days. While funds for religious performances are more than adequate, the number of religious practitioners has shrunk. The old custom according to which every middle son should become a monk and every girl who is the second of three sisters a nun is no longer observed, and in 1976 there were only five monks and three nuns attached to the Marpha *gompa*. Yet, the Buddhist faith is firmly rooted in the community, and there are chapels in private houses whose owners faithfully perform morning and evening the prescribed daily acts of worship. In this respect too there is a great difference between the vitality of Buddhist practice in Marpha and the lingering decline of religious life in Tukche. The resilience of Marpha as a Buddhist centre is demonstrated also by the fact that in recent years several young Gurungs have come to Marpha to study with the lamas of the *gompa*. Interest in Buddhism seems to be growing among the Gurungs of the villages north of Pokhara and several lamas of Thak Khola have gone there to minister to the Gurung devotees.

The description of social development in Marpha would be incomplete if the emigration of individuals and whole families to the middle ranges and the *tarai* were not taken into account. As in other high-altitude regions, a large part of the able-bodied and mobile population has always spent the winter months in areas with a milder climate, and this type of seasonal movement was combined with petty trade or the setting-up of temporary roadside inns (*bhatti*) where Marpha women catered to the wants of travellers, including the many Gurungs and Magars returning on leave from service in the Gurkha regiments of the British and Indian army. Between late November and March, close on three-quarters of the population of Marpha used to move south, and this seasonal migration followed a well-established routine, most families spending the winter months year after year in the same place. But in the old days they invariably returned to Marpha in time to participate in a festival which is accompanied by archery competitions. With the decline of Tibetan trade and the growth of commercial opportunities in the middle ranges and the *tarai*, more and more Marphalis extended their stay outside Thak Khola, and some bought houses and even land in the regions of Pokhara, Seti Khola, Andhi Khola, Tansing, Butwal and Bhairawa. While at first they used such establishments merely as winter quarters, some families developed them as their main domicile. By 1976 a substantial number of Marphalis had permanently settled in their second houses and rarely visited Marpha.

In 1962 I recorded the composition of 125 households of Marphalis resident in Marpha, and to these had to be added 20 Damai houses and 5 Kami houses, both untouchable Nepali-speaking artisan castes peripheral to the Marphali community. In 1976 only 40 of the erstwhile 125 Marphali households still had their main residence in Marpha, whereas 32 families had establishments in the lower country but continued to return to Thak Khola for the summer months and cultivate their land. No less than 47 families however had permanently settled outside Thak Khola, mainly in places such as Pokhara or Butwal, and some of these had sold their land in Marpha. Yet it was pointed out to me that their places had been taken by members of growing families, and that the village of Marpha, crowded together on a limited site surrounded on three sides by steep hill-slopes, could not accommodate all the Marphalis if those who had emigrated chose to return to their homes. There is no likelihood of such an occurrence, but even those who have settled outside Thak Khola seem to retain for the time being their identity as Marphalis. Marriages outside the community are still extremely rare, and the emigrants make conscious efforts to retain their Buddhist heritage even though they live largely among Hindus. In Kheireni, Marphalis have established a library of Tibetan sacred books housed in a place furnished as a small *gompa,* and from this any Marphali can borrow scriptures for domestic ritual performances. Lamas from Marpha and elsewhere are being employed to perform Buddhist rites, and if a Marphali dies in Pokhara or elsewhere in the lower country, the body is cremated and a small piece of bone is brought to Marpha within 49 days. With this, part of the traditional funeral ceremonies are performed, and the bone is ultimately deposited in the *khimi,* the clan shrine, a crude stone structure standing in a field. As long as the deceased has property in Marpha this ceremony is his due, but if he has sold his house and land, the dependents must pay a fee of Rs 50 to the *mukhya* for permission to deposit a piece of bone in the clan shrine.

The economic decline and partial depopulation of Tukche and other Thakali villages, and the continued prosperity and vitality of Marpha, Shyang, Thine and Jomosom, has created a new situation in the relations between the Thakalis proper and the Marphalis and other Panchgaonlis. As long as the leading Thakalis held important government contracts, such as the monopoly of the salt trade, and exercised administrative authority, expressed in their title *subba*, they were so obviously in a dominant position that the people of Panchgaon conceded their higher social status without question. A member of the most prominent Thakali family of Tukche discharged the function of *chikyap* in Marpha, and though his headship was mainly ceremonial he did at times act as broker between the villagers and government.

The Thakalis' social superiority in relation to the Panchgaonlis did not manifest itself in ordinary social intercourse. There was never a formal ban on interdining, and although some of the more superior Thakalis claimed to disdain the hospitality of Panchgaonlis on the ground that they ate yak meat, other Thakalis mixed freely with Panchgaonlis in eating and drinking. The distinction between the two groups was expressed mainly in their economic relations. Numerous men and women from Panchgaon used to serve in the Tukche houses, whereas no Thakali ever accepted employment in the house of a man of Panchgaon. To have done so would have lowered his status in the eyes of his own community, whereas conversely, a man of Marpha was not necessarily looked down upon because he had entered the service of a rich Thakali merchant.

The inequality between the two communities, which was emphasized by the Thakalis and not seriously contested by Panchgaonlis, was unlike the status differences between two Hindu castes, for it was neither institutionalized nor expressed in ritual behaviour. Today there seems to be less justification for such social inequality than ever, for the political influence which the Thakalis of Tukche used to exert over the villages of Panchgaon has disappeared, and a community such as Marpha is economically far securer than most Thakali villages. The time has long passed when young women from Panchgaon drifted into domestic service in Tukche, and men worked as the muleteers of Thakali masters.

Yet the Thakalis' sense of superiority in relation to Panchgaonlis has not yet been overcome, and even in Pokhara where Thakalis and Panchgaonlis trade on equal terms, the former have so far refused to admit Panchgaonlis to membership of the prestigious 'Thakali Social Reform Organization'. This may partly be due to the indisputable fact that some of the leading Thakali families have attained positions of great wealth and commercial prominence in Kathmandu and such industrial centres as Bhairawa, positions which cannot be matched by the economic achievements of any Panchgaonli. Inspired by the success of these early emigrants from Tukche, other Thakalis now living in Pokhara do not want to be identified with people from Panchgaon who have settled in the middle ranges only in recent years. Although even the richest Thakalis have not succeeded in gaining recognition as high-caste Hindus, they may feel that too close an association with Marphalis and other Panchgaonlis might prejudice the chances of any future attempt to raise their caste status. Marital unions of young Thakalis educated in Kathmandu with Chetris and Brahmans are still exceptional, but they suggest that the incipient break-up of caste barriers may ultimately benefit the social aspirations of the Thakalis.[1]

The choice open to both Thakalis and Panchgaonlis in the present climate of change will be discussed in the final sections together with the prospects for the future of all the populations of the Kali Gandaki region, for to a greater or lesser extent all these populations form part of an economic network which must be viewed in its entirety.

As one moves from Panchgaon northwards, leaving behind the district headquarters Jomosom, with its air-strip and modern administrative buildings, one enters a region of fundamentally different character. The landscape becomes more and more arid as one follows the course of the Kali Gandaki through the gap between the Annapurna and Dhaulagiri Himal. The fierce dry wind blowing daily through this gap from south to north desicates the centre of the valley and accounts largely for the stunted almost treeless vegetation; while the mountains to either side get more rainfall and hence provide pastures and some wood reserves. The villages lying on alluvial terraces above the banks of the river, and in the valleys of tributary streams are inhabited by a population appearing at first sight homogeneous and unmistakably Bhotia in appearance and culture. Rather illogically this region is known as Baragaon, though the number of villages is not twelve but eighteen, and it borders to the north on Lo (Mustang), to the east on Dolpo, and to the west on Nye-shang, more generally known as Manang. The region derives a sense of political unity from the fact that all its eighteen villages are comprised within a network which allocates to each village a definite place in a hierarchical system. There was moreover customary provision for gatherings of the representatives of all villages for the discussion of matters of common interest, but this has largely been superseded by the modern *panchayat* organization.

During the past 120 years or so the economic and political fortunes of the people of Baragaon have undergone a series of changes. Local traditions and the testimony of the ruins of substantial forts indicate that in the not too distant past chieftains drawn from a class akin to the aristocracy of Mustang ruled over the larger villages or groups of villages. The members of this class are known as Kutak, and there are still Kutak families in such villages as Dzarkot, Puta, Purang and Kagbeni. There are supposed originally to have been six 'houses' of Kutak, descended from six brothers who came from Dzarkot. While the eldest remained in Dzarkot, the other five established themselves in Dzong, Kagbeni, Puta, Samar and Dangardzong. When speaking Nepali the people of Baragaon refer to the Kutak as Thakuri, a term which in this context stands simply for any member of a ruling class and does not indicate any affinity to the Hindu caste of

[1]For a discussion of the Thakalis' reform movement aiming at their recognition as a Hindu caste of respectable status see my essay 'Caste Concepts and Status Distinctions in Buddhist Communities of Western Nepal', in C. von Fürer-Haimendorf, ed., *Caste and Kin in Nepal, India and Ceylon* (London), 1966, pp. 147–51.

Thakuris. Occasionally they are also referred to as Bista, but I was told that this was used only by Nepali speakers from Kathmandu who did not understand the local social system and terminology. As members of a martial class the Kutak were exempt from the payment of land revenue, in recognition of the assistance they had given to the rulers of Nepal in the wars against Tibet.

Nowadays only some main walls remain of the fort of the Kutak chieftains of Dzarkot, but the descendants of the former rulers still live in large and impressive houses. The best of these is built round a courtyard and contains a Buddhist chapel with a good collection of sacred scriptures. It is said that in the time of the great-grandfathers of the present owners, the Kutak of Dzarkot had to treat them with great deference and were expected to bare their heads the moment they entered the village gate. At that time the people of Baragaon looked down upon the Thakalis and Panchgaonlis, with the exception of the inhabitants of Thini, the seat of a legendary raja. But later this position was reversed, though there is no precise information on the causes of the Kutaks' fall from power. Most likely the local chieftains lost their influence when the rule of the house of Gorkha was extended to the whole of Thak Khola. Only the Raja of Mustang retained his autonomous status. Whatever the reason for the Kutaks' loss of influence may have been, the ascendancy of the Thakalis was the direct result of their acquisition of a customs contract which gave them control of the salt trade and cut out all other traders from this most profitable business. For the last three generations the leading Thakali families boasting of the title *subba* have certainly dominated Panchgaon.

Even when the salt monopoly was abolished in 1928, the hold of the Thakalis of Tukche over many Baragaonlis continued. In 1962 when I first visited Baragaon, many villagers were still indebted to Thakalis, and I heard many stories of their high-handedness; but also of some beneficial public works such as irrigation channels commissioned by Thakali patrons. At that time many of the people of Baragaon still lived as bond-servants in the great houses of Tukche, and Thakali creditors received regularly large quantities of agricultural produce as interest payment from Baragaonlis who had had to mortgage their land.

The departure of the great Thakali families from Tukche has brought about a drastic change in this situation. When they decided to transfer their business activities to Pokhara, Kathmandu or Bhairawa, they had no longer the determination or even the physical means to hold on to their assets in Baragaon. Some sold their holdings to local people on easy terms, while others found it no longer practicable to collect the interest on loans given to Baragaonlis, and in effect abandoned their investment which were in any case rapidly eroded by inflation. One of the Thakali merchants of Tukche had invested more than Rs 1,00,000 in villages in Baragaon, mainly by lending small sums to various people. But for the last six years he had not gone to these villages to collect the interest, and the debtors did not make any move to pay him. There is a general assumption that the debts have lapsed, and it might now be very difficult for a Thakali to collect interest or obtain repayment from Baragaonlis. The same Thakali who had invested Rs 1,00,000 in Baragaon was *chikyap* of the village of Tetang, but gave up this position in 1966, even before he ceased collecting his interest.

Thus the Bhotias of Baragaon have emerged from their subservience to the Thakalis of Tukche. Freed from the burden of old debts, they have greatly gained in self-confidence, and while in the past they often worked as carriers for Thakalis, they now trade on their own account.

However, their economy does not depend entirely or even mainly on trade, for agriculture and animal husbandry provide a substantial part of the food consumed in Baragaon. All land which can be irrigated is used for cultivation, and the Baragaonlis are skilled in the construction and maintenance of irrigation channels: there are special village officials whose duty it is to watch over the distribution of water. Yet, there are limits to the expansion of cultivation, for water is generally scarce, and in some villages not even all of the existing terraced fields can be cultivated.

One such village is Tetang, which lies in an arid side valley of the main Kali Gandaki valley. Tetang is something of a mystery. Throughout Thak Khola one encounters the belief that Tetang is very rich. It is said that at one time even the leading men of Tukche borrowed money from Tetang, and that some families have hoards of silver coins buried in the ground which they replenished generation after generation. Yet, the village which consists of three compact, fortress-like clusters, appears anything but affluent. The houses, built wall-to-wall, are up to five or six storeys high, but are narrow, dark and untidy, offering only very cramped living space. Being short of water though not of land, the Tetang people cannot grow enough grain to feed themselves throughout the year. To make good the shortfall some go to trade in India, others sell goats in Jomosom and many earn wages in kind by doing seasonal agricultural work, helping with the harvest in any of the villages between Marpha and Mustang. They used to take caravans of hundreds of goats north to Tibet and south to the middle ranges or even the *tarai*, but on account of the low level of the salt and grain trade, they have given up using goats as pack animals. On the other hand, in 1976 there were 70 mules in Tetang, and these were used to provide transport both inside and outside Thak Khola. Yet, the number of Tetang people who go on trading journeys is small. To the north they go only as far as Mustang where they barter home-grown barley for salt. Some income is obtained by weaving and selling blankets and shawls, but shortage of wool, previously imported from Tibet, sets narrow limits to the production of textiles. This is another example of the adverse effect of the Chinese restriction on exports to Nepal on the economy of the border people. There are about 1000 goats in Tetang, and each household sells annually an average of two to three goats for slaughter. Jomoson with its military garrison and several government offices now provides a ready market for livestock, and has in general replaced Tukche as a commercial centre. Some Tetang men also derive an income from hiring out their mule trains to traders from Jomoson, who have to bring up their supplies from Pokhara or Baglung. Others trade in India, and one rich man has specialized in the buying and selling of Tibetan antiques and curios, for which tourism has created a demand in Kathmandu and Pokhara.

Despite such rather exceptional enterprises of a novel nature, Tetang has not perceptibly changed since I stayed there in 1962. Indeed the conservative nature of Tetang society is striking. Within human memory no new house has been built in any of the three Tetang settlements, and the system of landholding and house-ownership is based on the assumption that brothers marry one wife and do not separate. If there are many sons in a family, one usually becomes a monk and one may marry a girl from a family without sons and go to live in his father-in-law's house. Only the eldest of several brothers has the right to inherited land, and is regarded as *dongba*—a term expressing full citizenship equivalent to the term *kuria* current among Thakalis. If one of the younger brothers does not get on with the joint wife and leaves the household, he forfeits the right to any share in the joint property and may either go to live in another village, such as Jomosom or Pagling, or join another house and cultivate a share.

Signs of dissatisfaction with the polyandrous family system are detectable among those men who have seen much of the outside world. Thus a man who has become wealthy by trading in India, recently bought a second house for Rs 1000 although he is the eldest son of his father and his only brother is still a young boy. It is believed that he will set up his brother in the second house, in order to avoid having to share his wife with him. His trading journeys have taken him as far as Gauhati and Tezpur in Assam, and it is clearly such adventurous entrepreneurs who are the initiators of change even within village society. Some other villages of Baragaon are altogether more outward-looking than Tetang, and have in recent years embarked on various new economic enterprises. Thus the village of Dzong has started a co-operative plantation of fruit trees obtained from the government agricultural farm at Marpha. A piece of fallow land was terraced and surrounded with a stone wall. An irrigation channel was then carved out of a hill slope to bring water to the orchard. The villagers gave their labour free for this ambitious project, and whenever the channel requires repair dozens of men and women turn out without delay to keep the water flowing. Here the new *panchayat* system has been taken up enthusiastically and at the time of my visit in 1976 labour gangs recruited from all of the villages of Baragaon were engaged in building a large *panchayat* hall in Purang. For this timber had to be brought from distant forests and there was great demand for unpaid labour. It was met by an arrangement by which every household had to provide one worker, man or woman, for a specified number of days, and this imposed a considerable burden on households with few able-bodied members. The people of Dzong were at the same time restoring one of their *gompa*, which was to be used both for ritual purposes and as a village hall. As a result of all these activities the men of Dzong had to devote much of their time to communal enterprises at the expense of their own agricultural work.

Notwithstanding the Baragaonlis' efforts in the spheres of agriculture and animal husbandry, they are not entirely self-sufficient in so far as food supplies are concerned, and obtain rice, pulses, chillies and tea either by barter or by purchase. Growing contact with the peoples of the lower country has moreover given them a taste for goods which can only be bought with cash. Kerosene oil, hurricane lamps, matches, electric torches, cigarettes, cotton cloth and various items of ready-to-wear men's clothing were beyond the aspirations of previous generations, but are nowadays found in the possession of many Baragaonlis. To earn the cash required to buy such goods men go on long trading journeys, and as Tibet is no longer open for trade in commodities other than salt and wool, they have had no other choice but to extend their commercial activities to the *tarai* and India. A totally unpredicted development in the trading pattern of Baragaon is the success of men from villages such as Dzong, Puta, Dzarkot and Purang in peddling textiles in India. The starting point of this enterprise may have been the sale of home-woven woollen blankets and scarves at a time when ample quantities of Tibetan wool could still be purchased by Nepalis. Today such supplies have been drastically reduced, but the Baragaonlis who had become familiar with the Indian market, did not accept defeat but started a totally different line. They bought machine-made sweaters and pullovers in Ludhiana in the Panjab and peddled then all over northern India. Trading on their unmistakably Bhotia appearance and dress, they easily convinced potential purchasers that their wares were the genuine product of Himalayan cottage industries and made of home-grown wool, while in fact they contained a large component of artificial fibre and were hence cheap when bought from the manufacturers in bulk. Having no overheads such as settled shop-keepers are bound to have, they were able to sell their goods at a large profit and to return to Baragaon with substantial sums of money. The success of the first entrepreneurs in this new field encouraged many Baragoanlis to follow their example, and today it is usual for men between the ages of eighteen and forty-five to spend several winter months in India and trade in machine-made sweaters. They usually take cash to Ludhiana and buy the textiles outright but men well-known to the manufacturers can also obtain goods on commission. A man setting out with Rs 5000 can make a clear profit of about Rs 3000 in the course of four to five months' trading and this enables a household to purchase all the ordinary necessities, and leaves sufficient cash for the payment of land revenue, minor fees and donations to the *gompa*.

In India Baragaonlis tend to avoid places such as Calcutta where they encounter competition from Tibetans, and concentrate on smaller towns and villages, and in particular on Assam. Many Baragaonlis are familiar with Gauhati, Shillong, Dibrugarh and Jorhat, but some have been as far as Kohima, North Lakhimpur, and Bombdi-La and their description of tribesmen such as Nagas, Daflas and Apa Tanis leaves no doubt that they have frequent contact with such tribesmen. In Dzarkot I met a young man of the Kutak class, who had just come to his village after spending three years in Nagaland where he had worked in the store of a kinsman now settled in Kohima. Such long sojourns in India are unusual, but the number of men engaged in the peddling of textiles and particularly of sweaters is large, and all seem to make good profits. While it is impossible to foretell how long Nepalis will be able to maintain this very rewarding trade, for the time being it fills very well the gap created by the decline in trade with Tibet. It seems that even the leading Kutak families do not think it below the dignity of their menfolk to peddle textiles in India, and this attitude is explicable in view of the high regard in which Bhotias of all classes have always held long-distance traders going with their yak caravans to Tibet.

It is only in some villages that objections to lengthy trading journeys to India have been raised. Thus, in Lubra which consists of a closely-knit community of twelve households all adhering to the Bonpo faith, the lamas of the local *gompa* recently ruled that for three years no one should spend long periods in India, because this prevented the young men from studying and participating in *gompa* ritual. Anyone disregarding this ruling made himself liable to a fine of Rs 100. Five men who went to trade in India in the winter of 1975–6 duly paid this fine, and the villagers expected that in the following year too some men would rather pay a fine than forego their trading profits. Yet the ban on long trading trips is a significant indication that the more conservative elements in Baragaon society do not consider the making of money by peddling textiles in India an undiluted blessing, and are of the opinion that Baragaonlis can maintain themselves adequately by exploiting local resources, supplemented perhaps by the yield from business deals within the traditional trading areas of Baragaon, Thak Khola, Dolpo and Mustang.

There has however for many years been the practice of purchasing *dzoppa* from Sherpa breeders of Solu and Khumbu. The acquired animals, many half-grown calves, used to be driven through Tibet, but nowadays they are brought by a route leading east–west through central Nepal. The Bonpo lamas do not seem to have raised any objection to this practice, and in 1975 three Lubra men went to Rasua Garhi to purchase 40 *dzoppa*. They bought altogether 40 *dzoppa* of an average age of three years; seven animals died on the way, but the surviving *dzoppa* were sold at a marginal profit within ten days of their arrival at Jomosom. Had there been no losses through illness, the operation would have resulted in a very substantial gain.

It is such ventures which enable the Lubra people to find the cash for the maintenance of their religious institutions. The village *gompa* has recently been renovated and some of the frescoes have been renewed by a painter from Tukche. Moreover two Lubra youths were sent for study to a Bonpo centre in Himachal Pradesh at considerable cost to the community. Through them large numbers of printed Bonpo scriptures have been added to the small library. Though a minority faith within the context of Baragaon, the Bonpo religion is thus surviving, and it received a strong impetus when in 1973 a reincarnate Bonpo lama, born in India of Tibetan parents, came to Lubra and stayed there for six months teaching some of the local people. A few hours walk from Lubra a new Bonpo *gompa* has been built just outside Dzarkot, as part of the residence of a Bonpo lama who had spent Rs 10,000 on the construction of his new house and *gompa*.

The maintenance of *gompa* and the service due to the divinities represented by the images is a considerable drain on the resources of individuals who have inherited private chapels. Thus one of the Kutak families of Dzarkot owns a large house, comparable in the wealth and quality of wood carvings to the great houses of Thakali *subba*, and this contains a splendidly decorated and furnished chapel (*lha-khang*). In this chapel there are three gilded images as well as a complete set of the *Kengyur,* the Buddhist canon in 108 volumes, each carefully wrapped in red cloth. The images as well as *Kengyur* were brought from Tibet by the present owner's grandfather. The owner's son, a man of about thirty, told me that only by trading in India could he maintain this establishment. He had to light two lamps morning and evening, and for this required an annual supply of four tins of oil which at current prices amounted to Rs 640. Yet, to neglect the service due to images might result in misfortune or illness for the whole family. The thought that the descendants of aristocratic families preserving an ancient cultural tradition are now reduced to the status of travelling salesmen peddling sweaters in Assam may seem depressing to the outside observer, but flexibility is one of the virtues of the Bhotias, and the people of Dzarkot see nothing demeaning in such trade.

Intensified contact with the people of the middle ranges and lowlands has not only transformed the pattern of consumption and trade, but has also brought about a change in the self-image of the Baragaonlis. Until one or two generations ago the people of Baragaon were content to be regarded as Bhotias and identified themselves largely with the Tibetan-speaking populations of Mustang and Dolpo. Today, however, most of them call themselves 'Gurung', for they believe that it is to their advantage to identify themselves with one of the major tribes of Nepal. In doing so they follow a practice which has become common among several ethnic communities. The Bhotias of Mugu now describe themselves as Chetris and those of Humla want to be referred to as Tamang. The Thakalis' unsuccessful as-

piration to be recognized as Thakuris has already been mentioned as well as
the Baragaonlis' equation of Kutak with Thakuri. But while the latter
practice is only a device of classification comprehensible by Nepali-speaking
Hindus, the fashion of calling themselves Gurung signifies a conscious shift
in ethnic identification. When one asks a Baragaonli to which Gurung clan
he belongs he is invariably unable to reply, and few Baragaonlis have any
knowledge of the Gurung clan system. Yet, most villagers insist that they
are Gurungs, though in 1962 I still met old men who described themselves
quite frankly as Bhotias and mildly ridiculed their juniors' claim to be Gurung.
The adoption of the total name Gurung has created a number of inconsis-
tencies of which the Baragaonlis are not entirely aware. For the members
of the Kutak families, who have close kinship relations with the nobility
of Mustang, do not describe themselves as Gurung, though they are clearly
an essential component of Baragaon society. It is mainly the members of a
class known in the local language as Padungu who call themselves Gurung
when they speak to outsiders. Padungu, who form the majority of the population of Baragaon, stand midway between the Kutak nobles and the low-
status artisans. Yet, there are status distinctions even within the Padungu
class, and these distinctions are linked with the different ranking of the
eighteen villages of Baragaon. There is a group of five villages which rank
the lowest, and it is significant that these five villages speak a language
closely akin to Thakali whereas the higher-ranking villages speak dialects
closely resembling the language of Mustang. The highest status is enjoyed by
those villages which are known to have been founded by Tibetan lamas,
and this fact too conflicts with the idea that all Padungu are in reality
Gurung.

Yet the Baragaonlis' present claim to recognition as Gurungs bears out the
fact that ethnic designations are no more static than are social relations.
Historical studies must take into account this impermanence of tribal names;
for over two or three generations a population remaining basically the same
may appear affiliated to quite different ethnic groups, simply by a change
of designation.

The fortuitous nature of ethnic classifications becomes apparent in the
villages immediately north of Baragaon. These border on the domain of the
Raja of Mustang, and lying at a greater altitude than Baragaon, conform
in many respects to the general character of the villages of Mustang. There
too the inhabitants have begun to call themselves Gurung. Thus in Ghiling,
a place only two or three hours journey from the Mustang village of Gemi,
I was told that the inhabitants were all Gurung, but the men who described
themselves as such added that further north i.e. just across the Mustang
border, similar people were called Phalwa of Loba.

Ghiling, a village of some thirty houses, is virtually self-sufficient in grain.
Even though only one crop per year can be grown at this altitude, enough
wheat, barley and buckwheat are reaped to feed the villagers throughout the

year. Barley used to be exchanged for salt in Tibet, and since this barter-trade declined, the villagers' standard of living has fallen. Some of them still go to Tibet, but can only buy modest quantities of salt both for their own consumption and for re-sale. They take such salt to Dana and Tatopani, and exchange it for rice at a rate of two measures of salt for one measure of rice. A few men also go to India and peddle sweaters bought in Ludhiana, but such ventures are far less common than among the people of Baragaon.

The domain of the Raja of Mustang which begins a few kilometres north of Ghiling, differs from Baragaon both topographically and politically. Lying north of the Himalayan main range and jutting like a peninsula into Chinese territory, it forms part of the arid Tibetan plateau, where the oases of villages are separated by large stretches of desert-like country. Throughout the Rana period (1856–1951) Mustang was virtually a state within the state, and the Raja of Mustang continued to be recognized as a feudatory prince, whose position *vis-a-vis* the government of Nepal was comparable to that of the Indian princes *vis-a-vis* the British crown. But owing to the difficulty of communications and the absence of a political service, such as had ensured the Viceroy's overall control over the princes in India, the Raja of Mustang exercised unlimited authority over his subjects.

De jure the Raja of Mustang has no longer any special position, for his domain is supposed to be administered like any other part of Nepal. In practice, however, he still wields great influence, and the change-over from the traditional autocratic rule to the modern system of *panchayati raj* has not brought about any dramatic developments. The Rani of Mustang, a member of the Lhasa aristocracy, functions as *pradhan panch,* and several of the Raja's kinsmen continue to act as village headmen and have been elected *panchayat* members. It seems, moreover, that the *panchayat* under the chairmanship of the Rani meets only very rarely, and exerts little influence on the conduct of affairs.

The ancient capital of Mustang, known locally as Lo Manthang, is a walled city nearly square in shape. Inside the high walls, houses built of unburnt clay are crowded together in a labyrinth of narrow lanes. Most are double-storeyed, but the Raja's palace is a massive building four storeys high, matched in height only by one of the *gompa*. The Raja and his family dwell on the fourth floor, and it is there that he receives visitors and petitioners. The lower floors are used as store-rooms and for gatherings of the townspeople on festive occasions. Great quantities of seed grain are stored in this place, and at the time of the first sowing most of the townspeople obtain loans of seed grain from this store. Thus the palace is not only the residence of the Raja but also fulfils the role of a central granary serving the needs of the residents of Lo Manthang. The present Raja's father used normally to live in a smaller house in the nearly village of Tengar but when he died the people of the capital begged his successor to return to the royal palace and live among them, a request which he wisely granted.

The outward pattern of the traditional social order of Lo Manthang has thus been preserved, but the orientation of the inhabitants towards the outside world has changed. Until the early 1960's they were largely oriented towards Tibet, and Nepalese cultural influence was minimal. The Raja and his family had always intermarried with the Tibetan aristocracy and he as well as his courtiers had lived entirely in the Tibetan style. The father of the present Raja spoke only Tibetan, and as late as 1962 it was difficult to find in the villages of Mustang people who could express themselves in Nepali. One of the reasons for this isolation from the centres of Nepalese civilization was the special relationship between the people of Mustang and the Thakali merchants of Tukche. When Thakalis acquired the customs contract which gave them control over the salt trade along the Kali Gandaki route, they arranged for all the salt brought by the Mustang people from Tibet to be bartered for grain in Tukche, and hence there was no need for people from Mustang to go any further south. The Thakali merchants who at that time were fluent in Tibetan, acted as middlemen between the Mustang traders and the people of the lower country, and the great entrepôt of Tukche functioned in fact as a barrier beyond which caravans from Mustang did not go. The members of the Thakali *subba* family were strengthening their relations with the Raja of Mustang, and the most prominent member of this family succeeded in concluding a pact of ritual friendship with the Raja. The great financial resources of the Thakali *subba* had enabled them to gain economic influence in Mustang, and many of the people of Mustang became indebted to them in the same way as many of the Baragaonlis had become economically dependent on Thakali traders. Such contact with the Thakalis, however, did not impinge on the Tibetan character of Mustang society, for the Thakalis appeared Tibetan in language, dress and habits when they interacted with the people to the north, and assumed a Nepali guise in their dealings with the Hindu villagers of the middle ranges and the townspeople of Kathmandu. It was in *their* society that the two traditions clashed, whereas little of Nepalese civilization filtered through Tukche to the people of Mustang.

Today Mustang remains Tibetan in language and Buddhist in culture, and there is no other region in the Kingdom of Nepal where Buddhism flourishes as visibly as in the domain of the Lo Gyelbu, as the Raja is known to his subjects. Within the walled town of Lo Manthang there are three main *gompa* as well as an exquisitely decorated *lha-khang* in the palace. The oldest of the *gompa* (Tukchin *gompa*), supposed to be 500–600 years old, is an imposing building with six rows of wooden pillars, which must have been brought from a great distance as no suitable trees grow within a wide radius. This as well as the equally large Dzhyamba *gompa* contains images of both Sakhyapa and Nyingmapa divinities, and the two sects seem to have been represented in Lo Manthang for a long time. The essential daily services are performed in both these *gompa*, but the main focus of the religious life

of the town is Chyoti *gompa,* situated within a series of courtyards also containing the quarters of a large number of monks. It is believed to be 275
years old and is decorated with frescoes and sumptuously furnished with
large painted scrolls of exquisite quality. The most valuable of the gilded
statues on the altar are behind glass, and block-prints of a set of the *Kengyur*
rest in the library racks. A handwritten *Kengyur,* also belonging to the Chyoti
gompa, is kept in the Raja's palace for security.

While the townspeople take turns in the performance of the daily ritual
in the two older *gompa,* 55 monks are attached to the Chyoti *gompa,* and
it is there that all the major rites and ceremonies take place. The community
of monks under a head lama referred to as *khembo,* is the largest religious
community in the whole region, and there seems to be no difficulty over the
recruitment of novices. The number of nuns, which at one time was also
substantial, has shrunk however, and now only eight are attached to the
gompa. Some years ago the *khembo* suspended the rule that the second of
every three daughters of a family should become a nun, for many nuns used
to withdraw from the *gompa,* and he argued that it was pointless to force
girls to become nuns if later they wanted to marry.

Gompa staffed by monks are also found in many of the villages of Mustang, the biggest and most impressive being that of Tsarang. Most are kept
in good repair, and there can be no doubt that throughout the Raja's former
domain, Tibetan Buddhism is still unchallenged and very much alive. Yet,
the break of the cultural links with Tibet has an adverse effect on the religious
institutions of Mustang, for young monks can no longer go to Tibetan monasteries for higher studies and there is no more interchange with more senior
lamas.

While previously the study of Tibetan scriptures was the only form of
education available to the people of Mustang, today government schools
operate in most villages. The medium of instruction in these schools is
Nepali and many young people become fluent in this language. Since trade
with Tibet, though not completely interrupted, has lost much of its importance, and personal contracts with Tibetans inside Tibet have been made
impossible, the people of Mustang are now looking towards their Nepali-
speaking co-citizens for both trade and educational advancement. Members
of the nobility have begun to send their sons for higher education to schools
in Jomosom and Pokhara, and there are already some young men who have
a reasonable knowledge of English, and are familiar with Kathmandu as
well as with some Indian towns.

This re-orientation towards the middle ranges of Nepal is only beginning,
but it is fostered by both political and economic factors. The establishment
of the headquarters of the Zonal Commissioner at Jomosom has brought
Mustang within reach of the Nepalese government, and several visits of
King Birendra and other prominent personages to Mustang by helicopter
have given the Raja as well as the people of Mustang the feeling that the

King and his government are concerned with their fortunes.

While in the old days they could take their animals to winter pastures in Tibet, they now move with some of their herds as far south as Pokhara when there is no grazing in Mustang. Thus contacts with the Nepali people of the middle ranges are intensified, and there is a growing interchange of trade goods. People from Marpha and Jomosom come to Mustang and sell *rakshi,* cigarettes, matches, and other commodities, such as are found in the small shops of Thak Khola, and receive in payment wheat, mustard seed, medicinal herbs or cash. Men from Mustang also go to Pokhara to sell or barter medicinal herbs, which can be exchanged with rice at par. Most goods are transported on the backs of yak, *dzoppa,* mules, ponies and donkeys, but for the transport of salt from Tibet to Mustang goats are also still used, while this form of transport has become obsolete in Thak Khola.

The influence of the government agricultural farm at Marpha extends as far as Mustang, and the Raja has made a beginning with fruit-growing by purchasing saplings of apple, peach and apricot trees, as well as grapevines. These have been planted in orchards at Lo Manthang as well as at Gemi, and the prospects for fruit-growing seem to be good.

Although people in Mustang as elsewhere complain about a drop in their prosperity caused by the decline of trade with Tibet, there are no signs of real hardship, and the cultivation of grain crops, which depends entirely on irrigation, seems to meet most of the population's needs. As of old, supernatural aid is invoked for good crops, and the first sowing of wheat at Lo Manthang is done in the presence of the Raja and accompanied by the incantations of a monk who performs a simple sacrificial rite at the edge of the field.

On occasions such as this, the Raja acts as the representative of the whole community, and no one may start sowing until one of the Raja's fields has been sown. The charisma with which in popular belief the Raja is invested, spreads in a diluted form to all members of the Kutak class, the nobility of Mustang linked with the royal family by numerous marriage alliances. The number of Kutak families is small: among 200 *dongba* of Lo Manthang, i.e. household heads with citizen's rights, there are only fourteen Kutak families, and in all the other villages of the Raja's domain there are only ten Kutak households. The majority of the people are Loba (also known as Phalwa) and below them rank Gara artisans. The latter are not allowed to build houses inside the walls of Lo Manthang, and they are underprivileged in more than one way. Intermarriage between Kutak and Loba is considered irregular, and the offspring of such marriages is known as Shretsung. There is no way in which Shretsung can rise to Kutak status.

Both the Raja's and the other Kutaks' position is at present unassailable, and it seems that men of the Kutak class are regarded as the natural advisors of the Raja and leaders of village communities. A break-up of the social fabric such as can be observed among the Thakalis of Tukche is unlikely to occur in the foreseeable future, for the leading families of Mustang are firmly based on substantial landholdings, and unlike the Thakali traders they have no incentive to seek new outlets for their energies but can maintain and eventually improve their economic standards by developing the resources of their homeland.

CONCLUSIONS

This brings us to the problem of the causes and motivation of the social change we have observed among all the communities of the Kali Gandaki and Mustang regions. We have seen that the various ethnic groups have reacted in different ways to the changes in the economic climate brought about by external factors. These factors are firstly the restrictions on trade with Tibet subsequent to the Chinese takeover, and secondly the improvement in communication inside Nepal which have reduced the value of Tibetan salt and led to its replacement by supplies of cheap Indian salt. Whereas both these events inevitably reduced the profits which can be gained by long-distance caravan trade, they still leave the people of Thak Khola and Mustang several choices in the allocation of their resources in labour and capital. Social change, after all, is the sum total of the decisions made by a significant number of individuals regarding the manner in which they apply their energies to their best advantage.

The Thakalis of Tukche had been the first to make a conscious change in their aspirations and the way they applied their resources. Long before the profitable trade with Tibet had dried up, many of the richest and most successful merchants had transferred much of their wealth to places outside Thak Khola, and established business and industrial enterprises in Kathmandu, Pokhara and Bhairawa. The main motive was then the desire to increase their profits and to build up fortunes independent of business prospects at home in such places as Pokhara, against the need to uproot themselves and virtually abandon their palatial homes and an eminent social position in a relatively small community. Some of the Thakalis took this decisive step, while others tried to have the best of two worlds by maintaining two homes and living alternately in Tukche and rarely returning to Thak Khola.

When after 1960 Tukche lost its importance as an entrepôt for the salt and wool trade, those Thakalis who had retained their main domicile in Tukche were faced with the choice of following the example of earlier emigrants from Tukche, or accepting a temporary lowering of living standards and investing their resources in agricultural development including irrigation, an increase in yak herds and the provision of tourist facilities. In the event, most chose the former alternative with the result that Tukche has been shrinking and is on the way to a gradual decay. The choice of the leading men of Tukche was not whether to emigrate or remain in their traditional home. They had to choose between a relatively luxurious life in Kathmandu or Pokhara accompanied by a loss of any real political power, and a reduced standard of living in Thak Khola compensated by the retention of very considerable influence in local affairs. Within the present *panchayat* system a local power base is needed for any political advancement, and this is what Thakalis who moved out of their home district have virtually foregone. The value system of this ancient trading community clearly placed more emphasis on monetary gains than on the retention of political power, even though for about half-a-century the Thakali *subba* had combined administrative authority with the management of far-flung business interests.

The inhabitants of the other villages of Thaksatsae are more ambivalent in their reaction to the decline of the trade with Tibet. Being in possession of more and better land than the people of Tuckhe, they have more to lose by abandoning their farms and moving south, but lacking the stimulus of good trading prospects many young men nevertheless leave their villages for the Pokhara region or the *tarai* because they find life there more interesting and entertaining than in Thak Khola. The emigration of some families, moreover, has a snowballing effect on others. For the fewer people staying in Thaksatsae for most of the year the duller does life there become, and the greater the incentive to move also to one of the towns where Thakali communities are already thriving. Thus, one of the wealthiest and ablest young men of Kobang told me that during the previous winter he had been deadly bored, for he had no one to talk to, and rocks and trees had been his only company, a statement which seemed rather uncomplimentary considering that his wife had also remained at Kobang. The decision whether to move away from Thaksatsae is thus not made solely on economic grounds; the lure of the more exciting social life of a town with cinemas and other entertainments is thus a major factor in choosing between various courses of action.

The people of Marpha and other Panchgaon villages are faced with a choice similar to that confronting the Thakalis. So far the majority has decided on a different course of action. Equal, if not superior to the Thakalis in energy and enterprise, they too can make a good living in the commercial centres of the middle ranges, and some Marphalis have indeed moved to places such as Pokhara. Others, however, value membership of a closely-knit community and the pulsating social life of their home village more highly than prospects of greater financial profits elsewhere. Their answer to the present economic position is to intensify and develop the transport business which takes them to the middle ranges during part of the year, but also to seek new ways of utilizing the resources of their home village. Fruit-farming appears to them at present the most promising choice, and much hard work and capital is being invested in this novel enterprise brought to Thak Khola by an outside agency—namely the agricultural department of the government—and inspired by a dedicated Sherpa who had the right approach to the Marphalis, a people combining imagination with sound business sense. Both qualities are demonstrated by the further decision of some of the leading younger men to allocate resources to the creation of tourist facilities.

In Baragaon too, social change is largely the outcome of conscious decisions to embark on activities which had no place in the traditional economic system. The shortage of irrigation land as well as of pastures certainly sets natural limits to agricultural production, but there are several alternatives to the supplementation of income from farming. Faced with a similar situation, thousands of Jumlis migrate annually to India and seek employment as road labourers in Kumaun and Himachal Pradesh, while Gurungs and Magars traditionally enlist in the Gurkha regiments of the Indian and British armies. The men of Baragaon have chosen neither of these means to make ends meet. Apart from working at harvest time as seasonal labourers in Thak Khola, which is an old-established practice, they have hit upon the novel device of peddling textiles and specifically machine-made sweaters in India. The profits made from this activity are great, but so are the costs. It means living for months in a foreign environment and in considerable discomfort, and being separated from wives and children. Yet, on balance, the advantages outweigh the disadvantages, or at least seem to do so in the minds of the Baragaonlis. The fact that engaging in this trade is a free choice, and that work as a travelling salesman is not the only alternative to starvation, can be deduced from the situation in the village of Lubra. There the lamas have imposed a three-year ban on trading journeys to India, and it is unimaginable that men who are members of the village community would have taken this action if they knew that they condemned thereby their co-villagers to starvation or at least severe hardship. They must have thought that life could be carried on, though possibly with a lower level of creature comforts, even if no cash earnings from trade deals in India flowed into the coffers of many households. Similarly, some of the would-be traders took the individual and conscious decision to flout this ban and pay the penalty rather than abide by the lamas' ruling. Thus individual choice stands against individual choice, and the village community suffers neither from a suddenly-opening gap in its income, nor from a dramatic confrontation between traditionalists and innovators.

In Mustang, two events originating outside the local social scene forced upon the inhabitants a reassessment of their position and a reorientation of the economy. The Chinese threat to the border trade coincided by sheer chance more or less with the political restructuring of the political system of Nepal. Either event could have produced a number of different reactions.

The virtual abolition of the local monarchy and the inclusion of the Raja's former domain into a zone administered by a government official posted at Jomosom, could have been used to overthrow the hierarchic social order of Mustang and to do away with the privileges of the Kutak nobility.

Neither of these things has happened, and the people of Mustang continue to accept the authority of their Raja and pay him as much respect as they used to do when he still wielded absolute power. At the same time they have become aware of the superior power of King Birendra, who has paid several visits to Mustang, and have developed a new loyalty to the royal house of Nepal. As Tibet and the earlier influence of the great monastic centres of Buddhist civilization faded because of the changed political situation beyond the northern border, Nepalese influence increased and government schools established in all the major villages brought about a new cultural orientation. So far this influence has expressed itself through the spread of the Nepali language, now spoken fluently by the sons of the dominant families; but there are as yet no indications of an erosion of the interest in Buddhist rites and ceremonies, and the great annual festivals, such as the *tizi* (*dokyap*) are being performed in the traditional style.

These examples of the reaction of different ethnic groups to political events impinging on their traditional social and political structure suggest the great variety of adjustments to the novel situation. This adjustment may bring about a disintegration of the social order as it has done in Tuckhe, or it may induce a population to pull together and overcome difficulties by a heightened sense of co-operation and identity as in Marpha and Baragaon. Social change has progressed along various lines, but despite the constraints imposed by history and a harsh climate, alternatives between different courses of action have remained available throughout. There has been no attempt on the part of the government or any official agency to mould the process of change according to any preconceived pattern, and individual choices continue to be made in a situation of extreme cultural and social flexibility.

CHAPTER 6

SIKKIM

J. S. LALL

Sikkim offers a pre-eminent example of the operation of the processes of change in the Himalaya. Whether it is political structure, social organization, economic life or manners and customs, this small state has witnessed an unusual concentration of change in the past hundred years, and particularly since 1947. Its unique situation has exposed it to influences from four different quarters, playing upon an unusually diverse ethnic mix in its diminutive size of barely 7,296 sq. metres.

Hemmed in on the west by Nepal, on the north by Tibet and to the east by Bhutan and again by Tibet, the position of Sikkim is without parallel in the Himalaya. Surrounded thus, Sikkim enjoyed and still retains unusual distinctiveness. Though strongly influenced by Tibet in religion and customs, politically and economically the major factors for change have resulted from its close ties with India. It was a protectorate of India until 26 April 1975, when it became the twenty-second state of the Union of India.

HISTORY

The fascinating history of Sikkim must be passed over very briefly. It is only necessary to mention that the Lepchas are the indigenous inhabitants, and people of Tibetan origin called Bhutias[1] took refuge in the country

The author has made extensive use of his own notes and is indebted to Shri R. N. Haldipur (Principal Administrative Officer, 1963–9), for the loan of some of his. He also had discussions with Shri V. H. Coelho, Political Officer (1967–8), and would like to thank a number of un-named helpers without whose assistance this chapter could not have been written.

[1]See also the chapter on the Bhotiyas of Uttar Pradesh. The accepted difference in spelling has been retained. The derivation of Bhutia is from Bhote or Bhot which is itself derived from Bod, or Tibet. In a Sanskrit manuscript of the 7th century, Tibet is called Bhote. People of Tibetan stock who migrated to the southern face of the Himalaya are known as Bhotey or simply Bhutia. Communities of such people are found in a number of areas from as far west as Uttar Pradesh to Sikkim, even though their ethnic origins may differ. The Bhutias of Sikkim, for instance, are of Tibetan origin but it is not quite

after the schism in Tibet in the fifteenth and sixteenth centuries. One of their chieftains was crowned Chogyal, or religious and secular ruler, in 1642. Sikkim was quite an extensive country but it lost large areas to Nepal and Bhutan and finally to the British. It was the British presence in India which brought Sikkim into the arena of imperial interests in Asia.

Because of their concern with possible Russian ambitions in Tibet, the British decided to check Tibetan inroads and compel the ruler to return to Sikkim from the Chumbi valley in Tibet. A small expedition in 1888 achieved these aims. Under the Anglo-Chinese convention of 1890 the boundary between Sikkim and Tibet was fixed at the watershed of the river Tista. The Chumbi valley was thus recognized as a part of Tibet and Sikkim became a wholly Indian protectorate without any Tibetan attachment. The position of Sikkim as a British protectorate was accepted by the Lhasa convention of 1904.

Following the expedition of 1888, the British were virtually in control of the administration for the next thirty years. They decided to attract large numbers of Nepali immigrants to construct roads and extend agriculture. White, the first British Political Officer, wrote, 'But the country was very sparsely populated, and in order to bring more land under cultivation, it was necessary to encourage immigration, and this was done by giving land on favourable terms to Nepalis, who, since they knew it was to be had, came freely in.'

The encouragement given to Nepali settlers completely altered the ethnic composition of the country. When Sir Joseph Hooker visited Sikkim in the middle of the nineteenth century, he noted: 'Next to the Lepchas, the most numerous tribe in Sikkim is that of the Limbus' (Hooker, 1854, p. 137). By the time White left Sikkim he mentioned that there were about 6000 Lepchas, 6000 Bhutias and 50,000 Nepalis of an estimated population of 80,000. The Limbus were not even mentioned, though there are still a fair number in western Sikkim. The cumulative result of British policy was that by 1947 the Nepalis outnumbered the local people 2:1 in a total population of about 1,50,000. In 1977 they constituted 75 per cent of the population of about 2,03,000.

The strains inherent in this imbalanced ethnic mix stayed below the surface until the end of British rule in India for two main reasons. The first was the prestige of the protecting power and its representative in Sikkim to whom the ruler deferred for advice in almost all matters of any importance. In the second place, no one really questioned the benevolent rule of the

certain where the Bhotiyas of Uttar Pradesh originated. In Nepal, the name Sherpa means 'of the east' or 'people of the east'.

It should be mentioned that the name Tibet is of Tartar origin. Sherring explains: 'In the records of the Tarter Liaos in the 11th century the name is written T'u-pot'e, in which the latter syllable represents Bod.' The word Tibet is a European corruption of T'u-pot'e.' Sherring, 1906, p. 62.

Darbar, or Sikkim Government, which was essentially based on the personal rule of the Maharaja. However, Sikkim could not insulate itself against political forces originating from India. Political parties sprang up, one demanding merger with India, another insisting on transfer of power to the people and a third seeking to preserve Sikkim as a Bhutia–Lepcha homeland.

Initially, the party demanding transfer of power to the people was in the ascendent. Its leaders were given office but the actual exercise of authority exposed their immaturity and they had to be dismissed. In the ensuing frustration they turned against the Darbar itself. The tension which had been building up exploded in 1949; a small demonstration at the Palace, a minor event in itself, was the turning point. The ruler sought the help of the Government of India who lent a succession of officers to advise him on the governance of the State. The authority of the Darbar was gradually restored, and, in the changed mood, the ruler and the political leaders reached a compromise on parity in representation between the Bhutias–Lepchas and Nepalis. A constitution was framed on this basis which survived for over twenty years, mainly because it was underwritten by the Government of India which had a stake in the stability of Sikkim on account of its importance to national security.

The arrangement had much to recommend it for it recognized the special position of the Bhutia–Lepchas in Sikkim and the unique institution of the Chogyal who was both religious and secular ruler of the country. Modern political systems of election and responsible government were successfully grafted on to the traditional Sikkimese way of life. As such, it served as a remarkable model for change in the high Himalaya. This pragmatic adaptation of political institutions to traditional forms could have continued but for the general upheaval of 1974. Disorders in the State prompted the Chogyal to request the Government of India for help once again. Elections were held on the basis of one-man one-vote; the party demanding union with India and an end to the institution of Chogyal won a landslide victory. Its demand for merger was accepted by the Indian Parliament and on 26 April 1975, Sikkim became the 22nd State of the Indian Union.

It is uncertain whether the change represents an improvement on the regime which held sway during the previous twenty-five years. An enormously cumbersome bureaucracy and procedures, developed in an entirely different environment, have been superimposed on the older institutions which had been found adequate in the past. It is doubtful whether these changes have improved the quality of life. The experience only shows how important it is to reconcile the imperatives of change in the Himalaya with traditional institutions and the people's cultural values.

THE ETHNIC MIX

The census figures (based on Sinha, 1975, p. 10) for the last eighty years graphically illustrate demographic trends upto 1971:

1891	TOTAL	30,458
	Lepchas	5,762
	Bhutias	4,894
	Limbus	3,356
	Nepalis	15,458
1911	TOTAL	80,000
	Nepalis	50,000
	Lepchas, Bhutias and others	30,000
1931	TOTAL	109,808
	Lepchas	13,060
	Bhutias	11,955
	Nepalis	84,693

Census figures based on religious affiliations are (Indians and Tibetans excluded):

1951	TOTAL	137,725		
	Buddhists	39,395	Lepchas	13,625
			Bhutias	15,626
	Hindus including Limbus	97,863		
1961	TOTAL	162,189		
	Buddhists	49,894 (Lepchas & Bhutias)		
	Hindus	108,165		
1971	TOTAL	203,000		

Allowance must be made for a number of factors. In the early years of the census it must have been extremely difficult to cover the whole country effectively, particularly the more inaccessible areas. Statistics have traditionally been collected through village headmen or *mandals*. The vagaries of data collection in such conditions need not be stressed. Lastly, the floating or seasonal population, mostly of Nepali labourers and Tibetan muleteers, was often indistinguishable from the local population, particularly the former. Nonetheless it is possible to draw certain general conclusions. The statistics bring out, in the first place, the phenomenal increase in the number of Nepalis. In the forty years from 1891 to 1931 the increase was five-fold. This trend reflects the encouragement given to Nepali settlers in the first phase of the British protectorate. British attitudes changed after the first World War; increasing interest was shown in the Tibetan connection and the stream of new settlers diminished. Although it never totally stopped, the increase was thereafter also attributable to the high fertility rate amongst Nepali settlers and the practice of polygamy.

Figures for 1931 and 1951, which appear to be dependable, show a disturbing trend amongst the Lepchas of population stagnation; in certain areas it actually dropped. The increase in the ten-year period 1951–61 in respect of these two communities might be partly attributed to the presence of Tibetan refugees.

The most startling fact that emerges during the eighty years covered is the co nversion of Sikkim from a Bhutia–Lepcha country into a predominantly Nepali one. As we shall suggest in the succeeding section, while the Bhutias and Lepchas, for all practical purposes, have become a single composite community, the Nepalis preserve their separate identity. Consequently, the ethnic mix presents a picture at the present time of two broad heterogeneous groups in which the indigenous people are in the unhappy position of being heavily outnumbered by the immigrant Nepalis. The situation demands understanding and concern on the part of the Union and State Governments, which should be expressed in measures to safeguard the distinctive Sikkimese way of life which is threatened with extinction.

The Lepchas

The Lepchas are believed to be indigenous to Sikkim. They still call it Denjong or Land of Rice. According to Lepcha tradition they originated in Mayel, a legendary valley in the vicinity of Kanchenjunga[2], and have no tradition of migration. However, certain ethnic and linguistic similarities suggest that the Lepchas might have been part of the fifth century movement of mountain people from south-east Asia through Burma into present-day Arunachal Pradesh. Deflected by the Monpa areas at its western end, they appear to have found a vacuum in the secluded hills of Sikkim (White, 1909, p. 7; Coelho, p. 2; Sikkim Gazetteer, p. 27. The Lepchas, who were known as Rong-pa or valley folk, were fairly widely distributed throughout present-day Sikkim and Darjiling district in West Bengal; they seem to have been left to themselves until about the fifteenth century when, as stated earlier, the Bhutia influx from Tibet started.

The milder Lepchas gave way before the more assertive Bhutias. They lost the best lands and retreated to remote valleys and forest-clad mountains. About 2000 of them are now concentrated in Dzongu, an isolated tract in north central Sikkim, but Lepcha families and place names can still be found all over the state.

The economic relegation of the Lepchas was accentuated by Bhutia cultural domination. The Lepchas, who were spirit worshippers, embraced Lamaistic Buddhism somewhat half-heartedly; they never wholly lost their extraordinary understanding of nature. The womenfolk for example, make excellent nurses, and the men have been of immense help to botanists such as Joseph Hooker in locating and identifying species.

It is relatively easy even today to make out a Lepcha from a Bhutia. In 1952, Jawaharlal Nehru told the author that mist is good for the complexion;

---

[2]Five treasuries of great snow, 8582 meters at its unconquered summit. In deference to Sikkimese sentiment, the mountain being sacred, both the British expedition of 1955 and the Indian expedition of 1977 planted their flags a few metres below the summit.

the aptness of this remark is proved by the Lepchas whose complexions can best be described by comparison with the begonia, which is itself a shy species thriving in shade and moisture.

Bhutia cultural dominance induced a process of what might be called Tibetanization and the Lepchas generally sought assimilation, especially in court circles. Today many of the educated elite cannot even speak the Lepcha language. Intermarriage between the two communities is common and instances of Lepchas marrying Nepalese and Indians are not unusual, particularly amongst the Lepchas of western Sikkim and Darjiling. The distinctive Lepcha dress is now worn commonly only in Dzongu. The Governor's orderlies and the Palace guards have been put into Lepcha uniforms, which are picturesque and uncommon.

Noticeable differences can be observed in the pace of assimilation. It has hardly started in the Lepcha reserve of Dzongu while in eastern Sikkim it is advancing rapidly. Assimilation is most marked in western Sikkim, bordering Darjiling district, and Lepcha distinctiveness has all but disappeared in the older Indian district.

Special mention must be made of the Lepcha reserve of Dzongu. The area is cut off from the rest of Sikkim by the Tista river on one side and on the other by the Kanchenjunga massif. Nature, it seems, intended it as a refuge, and its natural isolation was accentuated by the ruler's decision to keep it as his private estate. Non-Lepchas were not allowed to settle there and Nepalis and Indians could not visit it except with the ruler's special permission. The Lepchas of Dzongu purchased their necessities from small bazaars at Mangan and Dickhu across the river. Such cereals and fruit as they needed were grown in their own fields. Paddy was grown in terraced fields but the other crops were planted on open hillsides, jhuming or firing of undergrowth being resorted to in cycles of eight years or even less. These practices have continued right up to the present day.

In Dzongu the Lepchas still observe some of the practices of their pre-Buddhist *mun* religion, the principal feature of which is a belief in spirits, combined with fairly elaborate rituals to control the evil ones and court the benevolent. Lamaism is practised side by side with *mun* and is rapidly replacing it.

The demographic consequence of the isolation of the Lepchas of Dzongu is that they have been diminishing in number over the years. In 1937, Gorer estimated that the population was 2000 in all, out of about 13,000 in Sikkim as a whole and 25,780 in both Sikkim and Darjiling district. Isolation has also resulted in accentuation of certain Lepcha characteristics, the most evident being non-aggressiveness. The forces making for change, for social adaptation and even minimal progress, were almost totally excluded. The recent decision of the Sikkim Government to open up Dzongu to outsiders is likely to be traumatic because of its suddenness. Had there been more intermixture in the past, by gradual stages, the Lepchas of this tract might have been able to adjust to outside society with less probability of damage to their own psyche.

The Bhutias

True to their Tibetan origin, the Bhutias left the hot and humid lower valleys to the Lepchas, and later to the Nepalis, devoting themselves more to trade than agriculture, as well as the vastly more rewarding occupation of being courtiers. Many of the leading Bhutia families followed the example of their ruler and married Tibetan wives. They wore Tibetan-style clothes and cultivated the refinements of their original home. The Bhutias never took to the tongue of the Rong-pa. Their spoken language is called Sikkimese; though it is regarded as a language on its own it is essentially a variant of Tibetan.

Gorer described the Bhutias as Sikkimese–Tibetans, a name reflecting their origin. It would, however, be truer to describe them as Sikkimese. If the Lepchas were Tibetanized, the Bhutias are not unaffected by three hundred years of separation from Tibet. To start with, the mild and damp climate of Sikkim has induced certain adjustments in a life style originally evolved to suit the severity of conditions in Tibet. Though at first they aped their ruler and tried to find brides in Tibet, intermarriage has become an increasingly frequent occurrence, particularly since the closure of Tibet. Marriages are restricted to immigrant Tibetans who have found homes in India and to the Bhutias and Lepchas of Sikkim.

The Sikkimese Bhutias have lost something of the swashbuckling manners of Kham, which was the original home of the ruling dynasty, acquiring inflections and nuances of dress, language and dance which are distinctively Sikkimese. The new year is celebrated on a different day and the great festival of Panglhapsol, accompanied by dancing and feasting, centres around worship of the Sikkimese deity, Khangchendzonga. Lamaistic Buddhism, it must be admitted, remained Tibetan in its religious essence. The great foundations of Pemayangtse and Tashiding and the ten other principal monasteries of Sikkim, are red sect institutions, with the usual complement of resident lamas, novices and lay members.

The Bhutia nobility were given the title of Kazis, borrowed, it seems, from India. Senior members of the Sikkimese nobility were accorded by consent the highest title of Athing, the last incumbents being the late respected Rhenock Kazi, and the learned Tashi Dahdul Densapa. The Kazis, apart from their local administrative functions, were the ruler's hereditary counsellors. Some of them held high office in the Darbar, or Sikkim Government, which derived its authority from the Chogyal. He was the fount of all authority, and the focus of Sikkimese identity.

No one with long experience of the Sikkimese elite can have failed to relish the sophistication and special flavour of their culture, which was reflected in varying degrees by all classes of Sikkimese. The composite Bhutia–Lepcha society was flexible enough to make adjustments arising from the intermediate position of Sikkim between Tibet and India. Exposure to both influences was an inevitable offshoot of the very considerable Indo-Tibetan trade through the country. They had become proficient in both Hindi and Nepali, although relatively few of them mastered the Hindi script. As already pointed out, the Lepcha reserve of Dzongu was a case apart. With the exception of Dzongu, social and cultural absorption of the Bhutias and Lepchas was virtually complete by the middle of the present century. Together, they are Sikkimese. It had become customary, for instance, for the Chogyal to refer to Sikkim as a Bhutia–Lepcha homeland. By and large the Sikkimese were responding well to the changed cultural and economic environment in the post-1950 period.

The Nepalis

It is the settlers from Nepal, coming in from the end of the nineteenth century, who have retained a cultural identity distinct from the racial amalgam which had become Sikkimese. The immigration policy of the British, and particularly the first political officer, has already been mentioned. Apart from the needs of road building and agricultural development, the British attempted to counterbalance the Tibetan inclinations of the ruling house by flooding the state with Nepali settlers. The writer of the *Sikkim Gazetteer* (1894) made a statement of British imperial interest that is unmatched for its sheer cynicism: 'Most of all will our position be strengthened by the change which is insensibly but steadily taking place in the composition of the population of Sikkim.... The influx of these hereditary enemies of Tibet [the Newars and Gorkhas] is our surest guarantee against a revival of Tibetan influence' (p. xxi).

The British had another motive. The predominantly Hindu Nepalis had already entered the Indian army in large numbers. Nepal was a staunch ally and the Nepali immigrants could thus be relied upon to create an interest in the Indian connection which would both neutralize the Sikkimese preference for Tibet and be welcomed by the Kingdom of Nepal. In 1908, when White left Sikkim, the Nepalis already outnumbered the Sikkimese. At that time there were about 50,000 of them and they called themselves Paharias, that is, of Pahar or Nepal, a proclamation of identity which is still widely prevalent.

Nepali immigration continued right up to about 1920, by which time the situation had changed. The threat of Russian expansion had receded and the British, through their political officers Sir Charles Bell and Sir Basil Gould, had established close and friendly relations with the thirteenth Dalai Lama. Thereafter there was a swing in the opposite direction—of interest in the Tibetan connection and consequent encouragement to the

Sikkimese. Nepali preponderance in Sikkim had, however, become an accomplished fact, and it was not possible to actively discriminate against Nepali immigration without damage to British relations with Nepal. Even though immigration was no longer encouraged, the Paharias kept coming.

With the exception of the tribals of western Sikkim, such as the Limbus and Tamangs, the Newars and a number of Nepali families who had settled in Sikkim for several generations, there was hardly any cultural assimilation, and ethnic assimilation none at all. The Nepalis remained a group apart from the Sikkimese. They dressed differently and spoke Gurkhali and Hindi. They never claimed to be Sikkimese, but if they resisted assimilation into a composite Bhutia–Lepcha–Nepali society, it could not be gainsaid that the prosperity of Sikkim rested to a large extent on their shoulders. They constituted the bulk of the peasantry and provided the administration and police with most of their strength.

To some Nepalis it seemed that parity with the Bhutia–Lepchas deprived them of their lawful rights under a democratic system of government. They did not quite see the Darbar's point that most of them were not Sikkimese. An attempt was made by the Sikkim State Congress to build a political party, as well as a heterogeneous society, based on parity between the three communities. This was a hopeful move for a people seeking to find a solution to ethnic diversity, but the party lost its moorings in the turbulent events of 1973–4. The Sikkim National Congress, which emerged victorious in the 1974 elections, is also a party of all the major ethnic groups; with wise management it could become the basis of a new all-embracing Sikkimese identity.[3]

The Northern Communities

Tucked away in northern Sikkim, at heights of nearly 3000 m, are the two highland communities of Lachen and Lachung. The Tista, which is *par excellence* the river of Sikkim, flows through Lachen, being known in this reach as the Lachenchhu. Thangu, the summer village, is at a height of about 4000 m and approximately 15 km upstream from Lachen. The stream diminishes in its higher reaches, taking an abrupt turn through the Giagong gap at the foot of Chomoyomo, one of Sikkim's most beautiful mountains, and then up to its true source beyond Tsolhamo. Here it converges with the head of the Lachung valley over the Donkya La pass at about 6000 m, below which the plateau first descends and then rises towards the border with Tibet.

On the Lachung side the valley descends to the pasture of Mome Samdong, where there are hot springs, then to the summer pasture of Yumthang and on to the main habitation at Lachung.

[3]It has since (1978) joined the Janata Party. In October 1979 the elections resulted in a new political kaleidoscope but parity was once again the basis of the majority party.

The two valleys are separated by high mountain features and lead more or less independent existences. Lachen is more fortunately endowed with pasturage. In addition to those at Thangu, there is the vast natural amphitheatre of the Lhonak valley rising from about 5000 to 6000 m, just below the Chorten Nyima Pass to Tibet and Jonsangla to Nepal. This secluded highland valley, unsurpassed in beauty, is ringed by peaks averaging about 7000 m, with the sapphire-blue gem of the Pema Tso, named after the princess of Sikkim, at its centre. The Lhonak valley was a favourite subject of the artist Maharaja, Sir Tashi Namgyal, whose work, in its latter phase, assumed a spirituality inspired by his fervent Buddhist beliefs.

Northern Sikkim commences at Dikchu in the middle Tista valley and includes the reserve of Dzongu which has already been described. Northern Sikkim thus comprises about a third of the territory of Sikkim, but the two northern communities with which we are immediately concerned occupy the valleys above Chungthang, the meeting-point of the Lachen chhu and Lachung chhu. The people of these highland communities range freely in this vast territory. Up to 1959 movement in the neighbouring areas of Tibet was also unrestricted. Before the winter snowfall they moved their yaks and sheep to safe pastures and came and went freely to such trade marts as Yatung in the Chumbi valley, Gyantse, Shekhar dzong and Tingkye dzong in Tibet. This free movement made it possible for them to pursue their traditional occupations of trade and animal husbandry.

Restrictions on movement began to be imposed from 1959 and the border was finally closed following the Sino-Indian hostilities of 1962. Since then the people of the northern communities have had to adjust to changed conditions both in their social habits and in their economic life. Marriage in Tibet is now a thing of the past. Instead, partners are found from the Bhutias of Sikkim and Tibetan refugees who have settled in India. In the course of time the Lachenpa and Lachungpa are likely to be increasingly Sikkimized but for the present they retain the characteristics which distinguish them from the other Bhutias of Sikkim.

NORTHERN ECONOMY

The people of Lachen and Lachung are essentially herdsmen and traders. The transborder trade was their mainstay. In the summer they crossed over into Tibet with textiles and a limited range of 'consumer goods' which found a ready market at that time in Tibet. In return they brought back wool, salt and borax. Many of them had traditional trading partners and, at the time of the closure of the border, twenty-seven Lachenpa and twenty-eight Lachungpa lost their investments in Tibet. Trade now occupies only a small place in their economy which has come to be based mostly on agriculture and horticulture, carriage of loads for government organizations and even labour on the roads. This has not entailed a perceptible change in their reliance on horse and mule transport and the breeding of these animals continues as in the past.

Certain traditional handicrafts, such as carpet-making and the weaving of blankets and tweeds, have had to be virtually abandoned because of the scarcity of Tibetan wool. Their own herds of sheep, never large, just suffice for their own needs. As a result reliance on agriculture and other occupations has become more pronounced.

At the time of the closure of the transborder trade, the Lachenpa and Lachungpa had fair-sized yak herds. Some were lost in Tibet and at the present time they have 690 and 840 respectively. Losses of sheep were even more severe; each of them lost over 4000 head. The two communities were almost wholly dependent on Tibet for grazing. Animals from Lachen spent seven months in Tibet and those from Lachung were seldom brought back even in winter. The loss of grazing facilities, which were traditional and for which levies were paid to the Tibetan authorities, is likely to result in serious overgrazing of the pastures in northern Sikkim. Though the Lachenpa are more favourably endowed in this respect, because of access to the Lhonak valley, even they have begun to feel the loss. The switch from pastoralism and transborder trade to more settled cultivation, mostly of buckwheat, barley, potatoes, and small-scale horticulture, is bound eventually to affect their life style.

Social Organization

There are 130 houses in Lachen and 150 in Lachung while the population figures are 850 and 950 respectively. The valleys are therefore by no means heavily populated and the people live in close proximity to each other in the headquarter villages. The phenomencn of summer and winter villages, which has been a feature of community organization amongst the Bhotiyas of Uttar Pradesh, is much less developed in Sikkim. While there are a few houses in Thongu, there is hardly any habitation to speak of in the higher reaches of the Lachung valley. There are at best some scattered sheep pens and graziers' huts. In the past, when herdsmen moved their animals to Tibet, they took their yak-hair tents with them and camped amongst their animals.

It is in the organization of their community life that the people of northern Sikkim are specially distinctive. Each community has an assembly called Dzumkha, composed of the heads of separate households. This assembly meets in a public hall known as *mong-khyim*. Once a year the Dzumkha elects two *pipon*s or headmen, and two *gyapon*s, who act as constables, messengers and odd-job men. It is the function of the *pipon*s to call the assembly to conduct public business. Most important decisions are taken in the village assembly, such as on grazing and cultivation programmes, seasonal movements of the community and disbursal of government assistance.

The Dzumkha is perhaps the most perfect form of democratic government anywhere in the country. They elect the *gen-me*[4], a body of respected elders who assist the *pipon*s, or executive heads, in settling disputes. Assisted by the *gyapon*s, the *pipon*s also collect government levies such as land revenue and forest tax, and generally represent the community in dealing with the government.

The monastery is the focus of community life to a greater extent than elsewhere in Sikkim. In this respect its importance resembles that of Tawang monastery in Arunachal. The lamas perform suitable *puja*s on all important occasions and the people have a genuine veneration for the forms of Mahayana Buddhism.

In the past the people of Lachen and Lachung lived very much their own lives, absorbed in their cycle of herding and trading. Since the closure of the border they have become much more dependent on the government at Gangtok; consequently their focus, while it will always remain within their community life, will be directed increasingly towards the assembly of the people of Sikkim as a whole.

These northern communities have shown remarkable resourcefulness in adapting to the cataclysmic change in their area, which differed from the normal processes of social change because it was induced by events outside their control and entirely external to them. Their strength lies in their community organization which has enabled them to face with conspicuous success the far-reaching changes of the last two decades.

THE MONASTERIES

In the early years of this century, when Sidkeong Tulku was the Maharaja, Sikkim was visited by an extraordinary French lady called Alexandra David-Neel. She spent thirteen years in Tibet and her writings introduced to the world much that is now known about lamaistic Buddhism in Tibet. About Sikkim, she said: 'The monks of Sikkim are for the most part illiterate and have no desire to be enlightened, even about the Buddhism which they profess. Nor, indeed, have they the necessary leisure. The *gompas* of Sikkim are poor, they have but a very small income and no rich benefactors' (p. 25).

Though lamaistic Buddhism continues to be the official religion, it is professed mainly by the Bhutias, Lepchas and Newars, along with a few of the other tribal groups such as Tamangs, and the Buddhistic overlay wears thin in Dzongu where *mun* traditions survive. One of the deeper changes that has taken place imperceptibly in the last fifty years or so is the transition to conditions in which reverence for the Buddhist religion is both genuine and widespread, even though it would now be incorrect to describe contemporary Sikkim as a theocracy.

[4]Body of elders : gen—elder: me-man

The twelve principal monasteries of Sikkim still receive government grants and there is an ecclesiastical department in the Sikkim Darbar. The influence of the monasteries has been diminishing, while fewer and fewer young boys are being sent by their families as novices for the priesthood. The last Chogyal, who is himself an incarnate lama deeply versed in Buddhist philosophy, was greatly concerned at this loss of interest. He accordingly set up a training school or *cheda,* where about forty novices and monks are being given new insights into the philosophy and practice of Buddhism. Fresh impetus in a different way is also being given to this Buddhist revival through the presence at Rumtek of the renowned teacher and mystic from Tibet, Karmapa Rimpoche. The small monastery on a spur facing Gangtok has become a place of pilgrimage for the devout from all over the world.

The establishment of the Namgyal institute of Tibetology at Deorali, a few miles from Gangtok, has also stimulated interest in Buddhist texts and philosophy. Scholars from all over the world have benefited from the resources of the institute and the opportunities for fellowship and exchange. Sikkim has consequently become an important centre for Buddhist studies.

THE ADMINISTRATION

Prior to 1949, a rudimentary form of government had been set up by the British. The Ruler was the repository of all authority. Certain departments had been set up in a secretariat in Gangtok, all of them answerable to the Maharaja. Apart from a police force and minimal departments of forests and public works, there was hardly anything resembling a local administration.

The British parcelled the country into twenty-four estates which were leased to the hereditary nobility, some early Nepali settlers and one Bihari. A fixed amount of *khazana,* or land tax, was levied on each lessee. What he himself collected through the village headmen, or *mandals,* was left to his own ingenuity. Though standard rates based on area and kind of land (wet or dry) were laid down by the State, in practice the lessee could settle as many newcomers as he wanted, from whom it was easy to extract initial payments in the form of *nazrana.*

The lessee also exercised some revenue and criminal jurisdiction. Personal differences were resolved by traditional courts and there was a chief magistrate in Gangtok to dispose of such cases as people took to him. The Ruler acted as a court of final appeal. The Lepcha reserve of Dzongu was administered directly by the Ruler's private office while the northern communities enjoyed a kind of cantonal autonomy.

It suited the British to invest this feudalistic system with a spurious permanence, on the model of the squire on the hill and the toiling peasants below. The latter were expected to render customary labour to their lessees and also for the benefit of the Ruler and political officer while they were on tour. The system so established inhibited development and had continued more or less unchanged until the mid 1940's. Its obvious anachronisms then led to resistance.

Opposition developed over *jharlangi*, or the hated impost of forced labour, and it was ultimately abolished. The people also ceased to have any confidence in the lessee's courts and their jurisdiction was terminated. As a result of the political ferment which was growing in the country, the peasantry started withholding payment of *khazana*. The lessees in turn fell into arrears. In 1949 it was found that the state's coffers were empty; hardly anything had been paid into them for two or three years, and the Darbar was eking it out on advances from the state bankers, a *marwari* firm in Gangtok bazaar. The writ of the Darbar scarcely ran outside Gangtok. In the country at large, the agrarian-cum-administrative system, if it can be called that at all, had broken down. The country was clearly ripe for far-reaching changes.

It was at this stage that the Maharaja asked the Government of India for the services of a succession of Indian officers to advise him and to administer the state. The secretariat presided over the negligible business of the Darbar which amounted to theoretical management of various departments ineffective in the country at large. One thing alone survived— the excellent system of noting in files inherited from British times.

It was thus urgently necessary to create some sort of administrative structure and to establish the authority of the Darbar in the countryside. Two *tehsils*, one for eastern and one for western Sikkim, were set up. Eventually a *sub-tehsil* was set up for northern Sikkim. *Tehsildars* were appointed who were given small police forces. Their first effort was to collect the outstanding *khazana*. Once this was done it could be said that an administrative structure existed. The *tehsildars* were invested with magisterial and revenue powers equivalent to those of an Assistant Collector in India. In earlier days the Criminal Procedure Code and the Indian Penal Code were used as guides; henceforth they were adopted. In civil matters the personal and customary laws of the parties were applied.

In 1950 the land leases were treated as having lapsed in consequence of the failure of the lessees to pay their land revenue, and all of them were abolished. The state thereupon entered into direct arrangements with the peasantry for payment of *khazana*. The *mandal*s were allowed to keep a small proportion for the trouble of collection, but by and large the revenue system as it emerged in 1950 closely resembled the *ryotwari* system in India. In addition, two operations of considerable importance were initiated in 1950—a land revenue settlement and forest survey. Each holding was measured and mapped and rights were inscribed in a permanent record. The *sukumbashi*s, or holders without title, were classed as secondary holders. The forest survey brought a vast potential source of wealth under planned management.

1966 saw the adoption of the Sikkimese Rural Indebtedness Act which had a significant influence on the socioeconomic life of the people. The traditional forms of indebtedness and pledging of land were made illegal and punishable with fines and imprisonment. Striking at the very roots of vested interests, the people burdened by debt and mortgage were released to participate in the new era of development.

In short, a direct form of administration was established which in some respects resembled the one which had prevailed in the non-regulation provinces of India. Simple and flexible rules were issued, and most of the administration was conducted through the *mandals* and other traditional agencies known to the people. The state government was taking steps to set up *panchayats* elected by the householders on a regional basis so that the *panchayats* would in turn elect representatives to the state assembly.

The Darbar's intention was that regional bodies would be based on the local *panchayats* and that ultimately the State Council itself would be returned by the regional tiers. However, the representational system which was eventually adopted for council elections was derived lock, stock and barrel from the one prevailing in India. Universal adult franchise was imposed on a traditional form of government for which it was not ready at that time. Extreme politicization, and some of its disturbing consequences, might have been avoided had institutions been built on traditional foundations.

The changes that were introduced in the representational system made too great a break with the past. Cataclysmic rather than gradual change became inevitable when competing interests endeavoured to manipulate the levers of political power. It would not be an over-simplification to say that, in the period 1940–70, the situation in Sikkim could be expressed in terms of a double dilemma: on the one hand the immobility of a feudalistic system provoked resistance and hence cataclysmic change; on the other hand the changes introduced were too extreme to be successfully grafted on to traditional institutions. The resulting political and institutional instability has yet to be resolved.

The administration has expanded considerably since the merger of the state with India. There is now a Governor, an elected Chief Minister and Council of Ministers and State Council. At the head of the secretariat is the Chief Secretary, supported by secretaries in the major departments. The *tehsildars* have become district officers, the police force has been expanded and placed under an Inspector-General of Police; there is a Chief Conservator and conservators of forests, and a supporting array of subordinate and field officers.

For a small state of just over 2,00,000 people, this administration is top-heavy and overelaborate. Specialized planning activity might well call for large numbers of technical and field personnel; but it is doubtful whether the new leviathan is justified in terms of public welfare.

THE ECONOMY

Present-day Sikkim consists almost entirely of the drainage of the river Tista and its confluents. Bounded on the west by the Singalila range and on the east by the Chola range, the lowest point is 246 m above sea level, at the confluence of the Tista and Rangit. From here it sweeps upwards to the summit of Kanchenjunga at 8582 m. Between these two altitudinal limits, Sikkim presents all the varieties of change that one sees in the Himalaya. Dense tropical rain-forests flourish up to about 1000 m. The species change gradually up to about 2000 m, and a distinct belt of bamboos of various kinds occurs between 2000 and 3000 m, merging with oak. The fir and rhododendron belt starts at about 3000 m, and continues up to just over 4000 m after which the mountains open into pastures, alpine flowers, dwarf rhododendrons and *juniperus squirmata*.

The soil cover throughout Sikkim, at least as far north as the Kanchenjunga massif, is generally unstable. Rainfall is heavy, being about 475 cm a year in Gangtok and over 600 cm at Karponang, only 16 km up the road towards Nathu La. This heavy precipitation occasions constant landslips up to altitudes of about 3000 m.

As may be expected, the vegetation changes dramatically in this varied scenario. Changes in the main flora have already been mentioned. Changes in cultivation are also sharply defined. Paddy is grown up to about 1600 m wherever water is available. Wet paddy is interspersed with maize in dry fields and the latter is grown even up to heights of 3000 m. In the wet middle-altitude valleys cardamom is grown and oranges have been planted extensively up to about 1500 m. The Lepcha reserve of Dzongu is a special case. Here *jhuming* (slash and burn) has continued right to the present and the main crops are cardamom, maize and millets. In the highlands, the traditional crops are maize, millets and, at high altitudes, buckwheat and a little barley. Apples were introduced in Lachen and Lachung by the British. Otherwise there have been very few changes in agricultural and horticultural practices.

Oranges and cardamom constitute the two principal cash crops. State revenue from them has risen from Rs 19,560 in 1950 to Rs 9,43,888.27 in 1975 without any appreciable extension of the planted area. The increase is accounted for mainly by the phenomenal rise in market prices. As a result, people growing these crops are now distinctly prosperous, even in the backward area of Dzongu. However, the Lepchas prefer to depend on Nepali labourers and they have thus opened themselves to exploitation, particularly in the sale of produce to Indian traders. The government has taken measures to restrict entry in order to protect them from the wiles of outsiders. The recent prosperity combined with improved communications, and establishment of schools and dispensaries, is rapidy changing the traditional life of Dzongu.

By and large agriculture in Sikkim is for subsistence. The average family-holding area is 3.5 ha and there are not many that are very much larger in size. Paddy yields are generally on the low side, averaging 14.25 maunds an acre. *Basti* rice, as it is called, may not appeal to the sophisticated but it has a distinctive and pleasant flavour.

Sikkim suffers from an overall deficit of food, aggravated by the presence of a large imported labour force for road construction. Rice has therefore to be imported before the monsoon and distributed to various centres. Oranges and cardamom are exported in the reverse direction. This purely Sikkimese trade has been virtually monopolized by Indian merchants. The overall picture is of an internal economy limited in scope and without much prospect of improvement. It is wholly dependent on the rest of the country for imported rice, sugar, oil, wheat, and the entire range of consumer goods.

Weekly bazaars scattered all over the state attract people from neighbouring hills and valleys. They bring with them their farm produce such as oranges, eggs, butter and cardamom. In need of money, the farmers generally sell off their produce at low prices, buy their modest consumer requirements and immediately repair to the liquor shops. They return home happy but not exactly affluent. Obligatory death ceremonies amongst Buddhists consume a large part of their earnings and also plunge them into debt for generations. What is called capital formation in advanced economies is practically non-existent. Indian traders and certain Nepali groups such as the Newars are probably the only people with ready money. Sikkim has a fairly typical hill economy, more isolated than most, dependent on India for food, consumer goods and a market for its produce. It had no separate economic existence even before merger and lacked the capacity to generate any significant economic advance.

But it is in respect of the Indo-Tibet trade which flourished from the Lhasa Convention of 1904 up to its final closure in 1962 that Sikkim played a major part as a channel of commerce. It would be no exaggeration to say that this trade through Sikkim was the most important single factor affecting the economy of the state. Under successive Indo-Sikkim treaties, Sikkim did not enjoy the right to impose duties or other imposts on the movement of goods. It was thus virtually incorporated in the Indian fiscal and economic system, but the benefit from the trade through Sikkim accrued to the Darbar and not to the people, from the monopoly enjoyed by the Sikkim State Transport Service in the carriage of goods up to Gangtok. Thereafter Tibetan muleteers took over the carriage business.

The trade itself was almost entirely in Indian hands. Though some Bhutias traded on a small scale, the people of Sikkim generally lacked the resources to compete with Indians from the plains, mostly of the Marwari community. It is much to be regretted that the British government did little to restrict Indian traders and nothing to ensure that the trade should remain in Sikkimese hands. The situation in this respect improved hardly at all after 1947.

With the closure of the Indo-Tibet trade, import–export and the transport business rapidly adjusted themselves to the changed situation.

DEVELOPMENT

The annual state revenue of about Rs 6 lakhs, heavily in arrears, was raised to about 40 lakhs in 1953–4 without any assistance from the Government of India. Thereafter an era of planned development revolutionized the economy of the state. The new emphasis on development was largely attributable to the keen interest of the Chogyal. The first Plan, introduced from 1954, resulted in an injection of about Rs 3.25 crores as aid from the Government of India. The last Plan of 1971–6 provided for an outlay of Rs. 20 crores, actual expenditure being as much as Rs 18.74 crores.

As may be expected, annual aid amounting to between Rs 3.25 and 4.5 crores has vastly strengthened the state's economy. Starting with projects such as small hydel schemes, demonstration and seed farms, and handicraft centres, it has been possible to provide for major schemes such as the Lagyap hydel project, a proposed pulp industry at Melli, and tea plantations for Tibetan settlers.

Soldiers returning home at the end of the 1940s gave a fresh impetus to the demand for schools and communications; thus the impact of the outside world was brought back to Sikkim by its own people. While those concerned with their own areas thought of modest change, the horizons of the people as a whole widened. The development strategy, however, advisedly concentrates on small-scale projects which benefit even relatively inaccessible areas. Considering that demand in Sikkim itself is very limited, these schemes will obviously have to be related to markets in the rest of India.

A start has also been made to preserve traditional skills such as carpet and blanket weaving and wood carving.

CHALLENGE AND RESPONSE

Most Himalayan communities are relatively coherent and occupy compact areas. Sikkim presents a case that is unique in three major respects. Firstly, though the territory is a clearly defined geographical unit, it extends through all three latitudinal belts—foothills, middle hills and high Himalaya. Secondly, the ethnic composition of its population is highly diverse, reflecting, partly, the diversity of its terrain, but more significantly the artificial encouragement given by the British, for their own imperial purposes, to settlers of a large group of outsiders. This group, the Nepali, has resisted absorption, and now outnumbers the composite Sikkimese three to one. The strains inherent in geographical diversity have been magnified by the disparate character of the two main ethnic groups, calling for a high degree of management skill by the state's administration.

A third factor is the variations in exposure to the processes of change. At one extreme is the constant movement of people, goods and services along the main trade and travel routes, reflected by the adaptability induced by these multiple processes. At the other, we see such secluded communities as the Lepchas of Dzongu and the tough northern communities of Lachen and Lachung, isolated, till recently, from the pressures which bring about socioeconomic change.

Finally, the entire state, with the possible exception of Dzongu, has been exposed to new phenomena resulting from the injection of very large development expenditure which is likely to accelerate the processes of change in Sikkim. Such concentrated exposure to change could induce unpredictable side effects, but the resilience which its people have shown so far is the surest guarantee of its future stability.

SELECT BIBLIOGRAPHY

COELHO, V. H. *Sikkim and Bhutan*. New Delhi. Indian Council for Cultural Relations.
DAVID-NEEL, ALEXANDRA. 1971. *Magic and Mystery in Tibet*. Corgi.
GAZETTEER OF SIKKIM, 1894. Reprinted 1972. New Delhi. Manjusri Publishing House.
GORER, GEOFFREY. 1938. *Himalayan Village: An Account of the Lepchas of Sikkim*. London. Michael Joseph.
HOOKER, JOSEPH DALTON. 1854. *Himalayan Travels*. London. John Murray.
MORRIS, JOHN. 1938. *Living with Lepchas*. London. William Heinemann.
SHERRING, CHARLES A. 1906. *Western Tibet and the British Borderland*. London. Edward Arnold.
SINHA, AWADHESH COOMAR. 1975. *Politics of Sikkim*. New Delhi. Thomson Press (India).
WHITE, J. CLAUDE. 1909. *Sikkim and Bhutan*. London. Edward Arnold.

CHAPTER 7

NEPAL HIMALAYA AND CHANGE

D. D. BHATT

The Nepal Himalaya stretch for over 800 km, forming one-third of the entire chain which extends from Afghanistan in the west to Burma in the east. They are broadly divided into four zones. The Sub-Himalayan or Siwalik Zone is a mid-Miocene deposit, 5–50 km wide, composed of sandstone, shale and boulders. The topography is rugged and its soils unstable. The Lesser Himalayan Zone, with a width of 85 km and an altitude of 3500 m has, as its most conspicuous feature, the Mahabharat *lekh* (mountain) with parent rock composed mainly of limestone, dolomite and quartz. Again, the topography is very rugged, with many east–west valleys and hills carved out by notable east–west alignments of three major groups of rivers— the Kosi, Gandaki and Karnali. It is in this region that sixty-two per cent of the people live. The Great Himalayan Zone has a width of 25 to 30 km and an average altitude of 7000 m. The major peaks of the Nepal Himalaya are found here, including Mount Everest (8814 m) called Chomolungma (Mother Goddess of Earth) in Tibetan. It is significant that of the 14 major peaks in the world, nine peaks belong to the Nepal Himalaya. They include, apart from Mount Everest, Kanchenjunga (8598 m), Makalu (8481 m), Cho Oyu (8153 m), Lhotse (8501 m), Gauri Shankar (7145 m), Dorje Lakpa (6989 m), Ganesh Himal (7406 m), Langtang (7246 m), Annapurna I (8090 m), Machhapuchhre (6997 m), Manaslu (8156 m), Dhaulagiri (8167 m), Api (7132 m) and Saipal (7035 m). In fact, in the Nepal Himalaya, there are 240 peaks of over 6500 m. Many major rivers cut V-shaped valleys through the main Himalayan chain, with gorges as deep as 1500 to 2500 m. Of these, the Kali Gandaki in Central Nepal is perhaps the deepest gorge in the world, 30 m wide and 5600 m deep. The Trans-Himalayan zone is called the Tibetan Marginal land and has an average width of 60 km and an altitude of 4500 m. Geologically, it is a geo-synclinical basin of the Tethys in which the sedimentation of the Palaeozoic era has accumulated.

The views expressed are personal and in no way reflect official opinion.

LAND USE AND DEMOGRAPHIC PATTERNS

Nepal is primarily an agricultural country, where the bulk of the population lives mainly in rural areas and the primary sector contributes a large share to the economy's GDP (6539 million rupees in 1969–70 at 1964–5 constant prices). There is hardly any urbanization or industrial growth. According to the 1971 census, the population of the country was 11.5 million with an annual growth rate of 2.07 per cent over the last decade. It is estimated that by 1980 the population of Nepal will exceed 13 million.

TABLE 15.1

Land Use Type

Land use type		Area in sq. km.	Percentage of the total
I **Agricultural**		19,800	13.98
Hilly region	5,795		4.09
Tarai region	14,005		9.89
II **Forest land**		44,750	31.61
Hilly region	28,750		20.31
Tarai region	16,000		11.30
III **Other Land**		77,027	54.41
Land reclaimable	18,600		13.14
Land un-reclaimable	26,441		18.68
Land under perpetual snow	21,121		14.92
Land under river-beds, canals and others	10,865		7.67

Source: HMG, Ministry of Food and Agriculture.

TABLE 15.2

Regional Variation

Criteria	Mountains and hills	Tarai and inner tarai	Kathmandu valley
1. Land area (%)	72.9	26.6	0.5
2. Cultivated land (1970/1) (in %)	26.9	70.7	2.4
3. Population density (sq. km)	58	128	1039

Source: HMG, Central Bureau of Statistics

Nearly 15 per cent of the land is under perpetual snow and another 19 per cent is categorized as non-reclaimable. According to more recent statistics, the forest area in the *tarai* and *bhabhar* has further shrunk to 22.2 per cent.

VEGETATION

The vegetation of any land is a reflection of the climatic factors. In Nepal, all principal climatic types and vegetation exist, from tropical, subtropical, subalpine to alpine, as well as arid highlands. Rainfall, in general, is higher in east Nepal than in the west while the southern flank of the Mahabharat *lekh* receives the full force of the monsoon. The foot-hills get more rain than the *tarai*; for example, at Dharan, which lies at the foot-hills, rainfall is 2336 mm (annually) while at Biratnagar, some 40 km away in the *tarai*, it is 1687 mm. Jumla, which lies in the rainshadow of the Himalaya, is comparatively dry; it receives a rainfall which varies between 491 mm to 926 mm.

In the south, which is an extension of the *cis*-Gangetic plain, the vegetation is typically tropical and subtropical. Formerly, this part was under dense *sal* forest, often intermixed with *Terminalia, Adina, Lagerstroemia, Eugenia,* etc. but much of this has been destroyed. Forest encroachment by settlers from the hills as well as from across the border has been fairly heavy in the eastern and western *tarai* districts. During the decade 1964-74, the government officially distributed 77,000 ha of *tarai* forest land to settlers from the hills. Encroachment in the *bhabhar* land, a narrow strip between the Siwalik (*churiya*) and the *tarai,* has accentuated the flood problem in the plains region below.

In the temperate forests, which lie between 2000 m and 3500 m, various kinds of oak, rhododendron, laurel, maple, fir, spruce and hemlock forests are found. Above 3500 m, *kharsu* oak (*Quercus semecarpifolia*) is quite dominant on south-facing slopes. This oak along with two other species, *Quercus incana* and *Quercus lanuginosa*, is widely used for fodder. Wherever oak forests have been destroyed, secondary scrub of *Symplocos, Vaccinium, Litsaea, Camellia* and *Daphniphyllum* (*D. himalayensis*) makes its appearance. In east Nepal, in the foot-hills and along shady banks of streams and rivers, tree ferns, *Pandanus* and *Cycas pectinata* grow profusely. Along the Sikkim border, up to about 2500 m, the *Quercus lamellosa* belt is succeeded by *Acer–Ilex–Magnolia–Osmanthes* association.

In the mixed broad-leaved forest, *Rhododendron arboreum* (*lali Iurans*, the National Flower of Nepal) is the dominant species. Often it forms part of vegetation which includes *Lyonia ovalifolia, Ilex dipyrena, Acer campbellii, A. sterculiaceum* and *Litsaea elongata. Daphne bholua,* the tree from which paper is made, is common in these forests but greatly exploited.

Fir and hemlock forests extend up to 3500 m, beyond which birch (*Betula utilis*) and several species of rhododendrons of many dwarf varieties, are the dominant form of vegetation. In the dry Himalayan valleys, dwarf plants, such as *Rhododendron setosum, R. lepidotum* and several kinds of rosaceous and leguminous plants along with *Juniperus recurva, Ephedra gerardiana, Lonicera spinosa* and *Cotoneaster microphylla* are the main scrub. The alpine meadows during spring are covered with a myriad species of plants, *viz.*

Primula, Meconopsis, Pedicularis, Iris, Artemesia, Saxifraga, Anemone, Allium, Fritillaria, Anaphalis, Potentilla, Gentiana, Fragaria, etc. The upper limit of plant life in the Himalaya extends well beyond 6500 m; *Saussurea gnaphaloides* was reportedly collected by Eric Shipton at 6500 m. In the bare morainic rocks and screes, at altitudes of 4500 to 5000 m, the vegetation consists mostly of grasses, and such plants as *Saussurea gossy-piphora, Rheum nobile, Eriophyton wallichianum, Potentilla cuneata* and some umbellifers. In Nepal more than three dozen rhododendrons have been collected, but a great array of species are found in the east, *viz. Rhododendron campanulatum, R. cinnabarinum, R. fulgens, R. wightii, R. thomsonii,* along with deciduous species like *Sorbus microphylla, Rosa sericea,* which cover extensive areas above the timber line.

According to forest statistics, in the hill region 27 per cent of the land is under crop plant and 58 per cent under forest. However, due to adverse topography, only 34 per cent has been categorized as commercial forest. Of the forest species of the hills, which are economically important, *champ* (*Michelia champaca*), walnut, chestnut, maple, magnolia, *tooni* (*Cedrela toona*) and *utis* (*Alnus nepalensis*), are notable. In the Karnali basin, in some pockets there are still good stands of cedar (*Cedrus deodara*). Blue pine (*Pinus wallichiana*) is a common species between 1500 m to 2500 m. *Larix* is a species of the dry Himalayan valleys. Orchids are common in the temperate forests; on mossy trunks of oak can be found species of *Plione,* while from stone ledges hang sprays of *Coelogyne.* One of the most magnificent orchids of the temperate forests is *Dendrobium densiflorum.* Another orchid which appears in late spring is *Rhynchostylis retusa,* whose pinkish-white racemes droop from *Celtis* trees in the Kathmandu Valley. *Cymbidium elegans,* which flowers in late October, bears pale white flowers in lax racemes.

Of the ground orchids, *Arundina graminifolia,* the bamboo orchid, *Gym-nadenia orchidis, Satyrium nepalense, Spiranthes sinensis, Calanthe tricarinata, C. plantaginea* and *C. mucuoa,* are all found growing in shady ravines or underneath rocks.

Medicinal Plants

Nepal's rich resources of medicinal plants are being over-exploited, with the result that production has declined over the years. Medicinal herbs, according to Dobremez (1976) comprise 3 per cent of the total exports of Nepal, valued at US $ 400,000 in 1974.

Nepalese herbs and drugs found a ready market not only in India but in such commercial centres as Singapore, Bangkok and Hong Kong. According to Dobremez, the Tibetan and Chinese pharmacopoeia have borrowed many plants from Nepal. Even before the tenth century, a sizeable quantity of them were exported to China.

Medicinal plants find prominent mention in the *Rigveda* and *Atharva Veda*. In the Hindu culture *charak* and *susruta* are identified with Ayurveda, which is the science of medicine. Due to varied climates and altitudes, Nepal produces a variety of medicinal plants. Nepalese herbal products are in demand in India, Belgium, Canada, Japan, the USSR and West Germany. Medicinal plants which are in considerable demand are: *Rubia cordifolia, Mahonia nepaulensis, Bergenia ligulata, Paris polyphylla, Valerians wallichii, Lycopodium calvatum, Swertia chairata, Aconitum napellus, Orchis latifolia, Rheum emodi, Picrorrhiza scropularise-folia, Ephedra gerardiana, Acorus calamus, Allium wallichii, Nardostachys jatamansi* (*iatamansi*), *Betula alnoides* (*Painyu*). While the actual collectors are not adequately compensated for all the trouble with which they collect the plants, the supply has dwindled, and the present regulatory measures are inadequate to control over-exploitation of medicinal plants.

Spices and condiments are also exported from Nepal. Ginger, turmeric, cardamom, cinnamon and chillies are exported and all these provide much-needed cash to the hill farmers. Efforts are being made to intensify their cultivation and use better methods of processing.

WILDLIFE

Nepal lies at the cross-roads of diverse faunal zones and there is very little endemism in the Himalayan fauna. In the Nepal Himalaya, Palearctic Mediterranean, Oriental and Sino-Japanese elements meet. According to Caughley (1969), the Himalayan–Mediterranean fauna is a minority conflux of species among the far more numerous Pan-oriental Indian and Indo-Chinese elements.

Wildlife throughout the midlands of Nepal has disappeared due to widespread deforestation. A visitor scarcely comes across wild animals between 2000 m to 3500 m. The barking deer, the white-crested *kalij* pheasant and the *chir* pheasant are scarcely seen. Between 2500 m and 4000 m, wherever forests have been spared, typical Himalayan fauna, such as snow leopard (*Panthera uncia*), blue sheep (*Pseudois nayaur*) *bharal* or *tahr* (*Hemitragus jemlaicus*), goral (*Nemorhaedus goral*), serow (*Capricornis sumatraensis*) musk deer (*Moschus moschiferus*), Tibetan wild dog (*Coun alpinus*), Pika mouse (*Ochotona roylei*) may occasionally be sighted. *Serow*, a goat antelope, and a shy animal, lives in dense bamboo jungles. Musk deer primarily inhabits birch–rhododendron forests at altitudes of 3500 to 4000 m. This is by far the most endangered animal, for a pod of musk, weighing about 12 gm, is worth as much as Rs 1000. There is some poaching across the borders into Tibet, where the Chinese are reported to be running a number of musk-deer farms. The poachers lay traps across a large area and both male and female deer are snared, a reprehensible practice, for it is only the males which yield musk pods.

Of Himalayan wildlife, the blue sheep, a protected animal, is widely found. Perhaps *Nayan* (*Ovis ammon hodgsoni*) and wild yak (*Bos grunniens*) also exist in the Nepal Himalaya, but there are no records to establish this.

In the Nepal Himalaya, ungulates can be seen grazing in the alpine meadows which extend up to the snowline. *Bharal* (*Hemitragus jemlaicus*) is the most hunted animal. It is supposed to be, according to Schaller, somewhat of an evolutionary link between the sheep and goat. It is an excellent climber and can be seen grazing on precipitous cliffs. During the winter it may descend to 2000 m where it is hunted. In summer, it moves to about 4000 m. A more common animal in the hemlock–fir forest is the Red Panda (*Ailurus fulgens*) which lives on bamboo. This animal is comparatively easy to see and trap, and there is a great demand for it in zoos abroad. Martens, weasels, civets and squirrels are other animals which inhabit the hilly regions of Nepal.

As a result of the initiative from FAO and WWF (World Wildlife Fund), Nepal has established a number of national parks and wildlife reserves in the mountains and plains region, offering protection to such animals as the one-horned rhino, tiger, wild buffalo, bison, black buck, swamp deer, pygmy hog, Gangetic Dolphin and Hispid Hare—all in the warmer parts of the country, the *tarai* and inner *tarai*. In the Himalayan region, there are the Everest National Park and Langtang National Park. Nepal's Tourism Master Plan lays considerable stress on the preservation of scenic areas as well as protection to endangered animals in the country.

The proposal to create Himalayan Shikar Reserves merits some consideration. It is proposed to set aside an area covering 8000 sq. km in Myagdi district (Dhorpatan) for blue sheep hunting. In this part of the country, there is an unusually high density of animals, 2–3 animals per square metre, and selective hunting in some areas where there is already some pressure from grazing, would be beneficial for the herd. But this sort of selective trophy hunting, where the hunter may have the choice of the biggest head, may upset the sex ratio. At present, selected organizations are reaping a rich profit from American hunters, who reportedly spend up to 10,000 dollars for a single head.

However, in the great stretches of the Himalayan region, wildlife provides the only source of animal protein, and is also as much a part of the economic activity as raising livestock or potatoes. In Manang, for example, the population of Gyansumdo supplements its income through the sale of musk, which Gurung (1976) says has helped to prevent a break-down of the local economy following the drying up of the trans-Himalayan trade in the early 1960's.

Regarding the Yeti, the 'abominable snowman' of the Himalaya, fact and fantasy intermingle. Claims have been made by members of mountaineering expeditions that very large foot-prints were sighted in the snow at altitudes above 6500 m, and a group of American naturalists recently asserted categorically in a personal interview that the tracks belong to an animal which is still unknown to science. Whether it may prove to be an example of an evolutionary link between the primates and homo sapiens, the Yeti continues to be part of the mythology of the Sherpas for whom it is as real as the Chomolungma.

BIRD LIFE

Bird life in the Nepal Himalaya is considered very rich. There are reported to be nearly 850 species of birds, both resident and migratory, which are adapted to different ecozones. Bird life in the hilly region of the country has, however, been depleted mainly due to deforestation. In the Mahabharat *lekh*, babblers, laughing thrushes, sibias, barbets, drongos, orioles, koels and yuhinas are easily spotted in the oak–laurel forest. The Spiny Babbler (*Acanthoptila nipalensis*) once considered rare, is a fairly common species in the secondary scrub forest of *Gaultheria. Symplocos, Myrsine* and *Camellia* are found between 1500 and 2000 m. The Red-billed Blue Magpie is a widely-found bird in the midlands, while the cuckoo or *koel's* 'kaphal pako' call is familiar to visitors during spring and early summer. This bird should not be confused with the *Himali Neuli,* which is the Great Himalayan Barbet (*Megalaima virens*). The Whistling Thrush, the *kalchaunre (Myiophoneus caeruleus)* is a common bird of the streams in the hills. In the Langtang National Park, more than 200 species of birds have been recorded. Besides the Red Start, which is found along the streams, choughs, wild crows, wagtails and snow pigeons are frequently sighted in the Upper Langtang Valley.

In the Nepal Himalaya different kinds of pheasants are seen. While the Red Jungle Fowl is a bird of lower altitudes, the *chilme,* Blood Pheasant (*Ithaginis cruentus*), is mainly a trans-Himalayan species founds between 3000 to 4000 m. The *monal,* Satyra Tragopan (*Tragopan satyra*) lives in thick jungles and is pre-eminently a Nepalese species. The Himalayan *monal* (*Lophophorus impejanus*), also called *danphe,* which is the national bird of Nepal, occurs right up to 4200 m. It is the highest-dwelling pheasant in Nepal. Its many-splendoured plumage makes it very attractive to poachers and it is an endangered species. When disturbed, it glides very gracefully down the slope, displaying its beautiful plumage. Also worth mentioning is the *chir* pheasant (*Catreus wallichii*), a western species, found in scrub jungles at altitudes of 1800 m to 2800 m. At one time it was reported to be of wider occurrence but now is found in a small area around Jumla. The Nepal *kalij* (*Lophura leucomelana*) a bird frequenting low altitudes (700 to 2600 m), has suffered most from hunters, but it still survives in small pockets in the foot-hills and in slightly larger numbers on the lower Himalayan slopes. This species is found throughout Nepal, but in the west intergrades with the Whitecrested *kalij* (*Lophura hamiltoni*) and in the east with the Blackbacked *kalij* (*L. melanota*). Besides shooting, which is often done at night with the help of torch lights, snares and traps are also used. A dozen small boys with snares can soon clear up all the *kalij* within range of a Gurung village (Roberts, 1977, p. 45).

Bird movements across the Himalaya is an annual feature; their routes lie along the river valleys, such as the Kosi and its tributaries. Fleming, Jr. and Sr., well-known ornithologists, have studied this phenomenon in Nepal. Their observations are worth mentioning here: 'Passing migrants include ducks, as reported by Hodgson, of which we have listed thirteen. Geese, cranes and even a pair of White Pelicans are on our list (Fleming, 1977, pp. 40-3). In central Nepal, in the vicinity of the Annapurna range, one can see hundreds and hundreds of eagles taking part in an annual migration. These include the Greater and Lesser Spotted Eagles and some Imperial Eagles. The Kosi River, behind the barrage at Hanuman Nagar, has turned into a huge mud flat, providing an ideal habitat for migratory and resident birds. In early spring, a visitor can observe thousands of Brahminy ducks and Bar-headed geese in flight. Besides these, there are many residential birds such as storks, cranes, darters and cormorants. A suggestion has been made to turn this into a bird reserve, which would be ideal for tourists. The area adjoins the Kosi Tappu Wildlife Reserve, of about 31.08 sq. km, which has been created for preserving a small herd of wild buffaloes.

Any change in the forest ecology also affects the bird population. Edward Cronin (1977, pp. 82-3) for example, noted in Dabhlay forest in east Nepal, that deforestation opened up the canopy, and dried the moist mat of mosses and other ground herbaceous vegetation. This adversely affected the Scimitar Babbler, whose long curved bill is designed to feed upon insects among moss.

Preservation of bird life is inextricably linked with the protection of forests in the hills and mountains of Nepal. The lack of enforcement of protective measures has adversely affected wildlife.

DEFORESTATION

In the past few decades in Nepal there has been heavy deforestation, both in the hills and in the *tarai,* which has brought about irreversible ecological changes. In the hills, the density of population is quite high—94 per sq. km, and as much as 92 per cent of agricultural households have less than one hectare of land. Due to population pressure, there has been loss in productivity, soil erosion, and poor management of farms, and forests of up to 3000 m have been cleared for cultivation throughout the midlands; only patches of scrub forest which are subjected to lopping and grazing remain. In Baitadi, which adjoins Pithoragarh district in Kumaun, India, a fine oak forest near Garhi has now been turned into a maize field.

Deforestation has progressed at an accelerated rate in the last three decades. According to Harka Guring, of the 638,300 acres of forest land in 1928, only 205,000 acres have been left in the eastern *tarai.* The rate of depletion has increased from 28 per cent during 1954–64 to 34 per cent during 1964–72.

In the last decade alone, nearly 30,000 acres of land have been cleared to make room for settlers in the two *tarai* districts of Kanchanpur and Bardiya.

SOIL EROSION

With limited cultivable land (16 per cent), any further expansion in the cultivable area has to take place at the expense of forest. In the hills and mountainous region, steep terrain and heavy rainfall during the monsoon, especially in the Mahabharat *lekh*, lead to the loss of top soil and hasten the process of soil erosion. As Erik Eckholm (1976, p. 78) has said: 'Topsoil washing down into India and Bangladesh is now Nepal's most precious export, but one for which it receives no compensation.'

In Nepal, Eckholm correctly points out, villagers must roam farther and farther from their homes to gather fodder and firewood, thus surrounding most villages with a widening circle of denuded hill sides.

Landslides in the hills of Nepal are as much due to the misuse of land by man, as to the general instability of the fractured and displaced rock formation. In the Trisuli watershed region, where a UNDP survey/demonstration project was carried out in the sixties, Tautsher reported that widespread erosion and deeply cut-in torrents have washed down large areas of newly-cultivated land, and greatly increased the gravel and silt load of the rivers several times. When silt load exceeds the transport capacity of the river and its tributaries, the excess water floods over and buries fertile tracts under silt. As the rainfall is mainly confined to the monsoon period, the run-off is much greater than the carrying capacity of the streams and rivers. In Nepal, a predominantly rice culture also encourages floods: on steep hill-sides where the dipping of the rock formation favours sliding, or where faults and cracks lead to deep penetration of the water, hill irrigation canals cross along counter-lines of very steep slopes, bringing the water to terraces. These canals channelize the flood water, which flows along the steepest slopes often causing deep gully erosion and landslides.

River beds in the *tarai* are rising several centimetres in a year due to silting. Much of this silt is washed down the Ganga to the Bay of Bengal, so much so that 'an immense new island' surfaced in the Bay of Bengal in 1974, all on account of 100,000 square metres of silt. One Nepalese expert, Dr G.L. Amatya, has estimated that Nepal is losing 164,000 cu. in. of top soil a year. The Karnali River is reported to move some 75 million cu. m of silt and debris every year, an amount that corresponds to 1.7 m soil cover of the whole Karnali Watershed (IBRD Report). Another problem associated with landslides is the damming of rivers. The Burhi Gandaki River was blocked for 29 hours in August 1968, resulting in the complete inundation of Arughat Bazaar. Similar incidents occurred at Kasauli Daretol caused by the Tinau River.

The rivers in Nepal, when they leave the foot-hills, change their course.

The Kosi river is gradually shifting its course westward, while the western rivers are being pushed eastward. Several rivers—Babai, Kankai and Narayani—have shifted their courses several kilometres in the past decades, a situation which may affect flood control and irrigation schemes in Nepal and also across the border in India.

HYDRO-ELECTRICAL POTENTIAL

Nepal's hydroelectric potential is estimated to be around 83 million kW, of which only 45,000 kW has been made use of. Several hydroelectric power stations have been planned, including Karnali which has a discharge of 9880 m³/second at Chisapani. Two big hydroelectric projects, Kankai (32,000 kW) and Kulekhani (60,000 kW) are under construction. Several microhydel schemes (up to 150 kW) are being constructed in the hill regions. These schemes have distinct advantages in Nepal where wood for fuel has become scarce, and in order to save forests from further denudation, alternative sources of fuel and fodder have to be found. Watershed management is the key to the future prosperity of the Nepalese midlands, but present human and capital resources make it impossible to tackle this problem.

ENERGY SITUATION

Wood is the principal source of energy in Nepal and comprises as much as 87 per cent of the total energy consumption. In the mountainous regions, intense solar radiation can be tapped for heating and running small power units. Experts believe that solar heating could satisfy 60–70 per cent of the heating requirements of many mountainous regions.

Though the energy requirement of hill farmers is limited to half-a-ton of dried wood per person per year—this itself puts pressure on the forest and creates energy shortages. In Nepal, the per capita energy consumption (kg of coal equivalent) in 1973 was 14, compared to 188 in India, 20 in Afghanistan and 29 in Bangladesh. There is more consumption of fuel in the hills (600/kg/man) than in the *tarai* (398/kg/man). The acute shortage of fuel in the hills is emphasised in the *Energy Report,* which points out that if the present rate of consumption continues, the whole of the commercial forest which comprises 12 per cent of the total, will be exhausted within twelve to fourteen years. In order to meet the high demand, a production forest area is needed of about 3.38 m. ha on which the best growing varieties with 8–10 years rotation will be planted. But at the present rate of effort and investment, this will take 135 plan periods.

The Energy Group has looked into the alternative energy sources, including cow dung, and has come to the conclusion that bio-gas could supply fuel for cooking to 50 per cent of the population of Nepal in 1975 (Table 15.3).

Though Nepal has a large number of cattle, each farmer has merely one or two, so that 60 per cent of the families do not have the economic potential for bio-gas. So far about 28 bio-gas plants have been installed in Kathmandu, Lalitpur, Bhaktapur, Palpa, Kaski and Sidhuli. Besides the scanty supply of substrate, which mainly consists of farm manure, as human faeces are not utilized, the low temperatures at high altitudes makes these plants comparatively unproductive. When the temperature falls below 12° C, fermentation does not take place. Furthermore, the initial cost of installing a plant—roughly Rs 400–600—is beyond the ordinary farmer. Community or cooperative bio-gas plants are not feasible in the Nepali social context. As regards heating fermenting vessels, this could be done by solar cookers, but again installation and maintenance would present difficulties.

TABLE 15.3

Theoretical Potential of Cattle Dung

Stock	Number (10^6)	Fresh dung (106^6 x MT)	Dung available for gas and fertilizer on dry basis
Adult cattle	3.876	12.597	1.8895
Adult buffalo	2.130	8.520	1.2780
Young cattle	2.584	4.264	0.6396
Young buffalo	1.420	2.343	0.3515
Total	10.010	27.724	4.1586

Source: 'Nepal—The Energy Sector'. T.U. Institute of Science.

FODDER

The question of fodder supply also requires consideration. The straw from paddy, wheat, millets, and other by-products, grass and leaves, form the fodder base in Nepal. Next to wood, husk, dung cake and vegetable waste form the bulk of the non-commercial fuel in the country (Table 15.4). In the hills, the use of dung as fuel is rather rare, though in the mountainous region, especially in the trans-Himalayan zone, yak dung cakes are used as fuel, which reflects the scarcity of wood for fuel in these areas. In the midlands, however, dung, *guintho*, is collected from the grazing grounds, fired into ash and then mixed thoroughly with grain before the latter is stored in bins. This is done to keep away the insects.

Of the several factors which trigger off large-scale migration to the plains, scarcity of fuelwood and fodder are the principal ones. *Lekh*s which used to provide fuelwood and fodder, have been exploited and now have little apart from *Ribes, Berberis, Crataegus* and *Sarcococca*. If it were not for the planted fodder trees, such as *Ficus, Celtis, Grewia, Rhus, Prunus, Melia, Morus,* etc., which sustain livestock during the lean months, life in the midlands would be impossible. In winter the farmers prepare a special gruel for cattle by boiling nettles with rice husk and a little salt and serve it every morning. It is called *kholvo* and is considered very nutritious.

TABLE 15.4

Non-commercial Energy Consumption by Type of Fuels
(1974/5)

Non-commercial fuels	Energy coal tonne	replacement %
Fuelwood	6,204,000	97.23
Husk	86,000	1.35
Dung cake	62,757	0.98
Total	6,380,757	100.00

Source: T.U. Institute of Science, 'Nepal—The Energy Sector'.

PALLACHAUNDLY VILLAGE: A MICROCOSM OF CHANGE

The typical nature of the problems facing the hills of Nepal can be illustrated from a village in the far west, Pallachaundly, which lies in Baitadi district. In 1973 the general economic condition was deteriorating, the whole of Baitadi district was in the grip of famine, the reasons being manifold: drought, low yield due to soil exhaustion, deforestation, and a rising population which in the past decade had grown at the rate of 1.7 per cent per annum. Meanwhile the size of farms had shrunk,[1] and so the only alternative for eking out a living was cultivating an *ijar*[2], and planting lentils and millet, usually, *kodo* (*Eleusine coracana Gaertn.*)

In Pallachaundly village, people complained that the productivity of land had gone down by as much as 50 per cent, and there was hardly a family which did not buy foodgrains. Earlier, only a few of the *mallo jat* (upper caste), families went for *besya,* the custom of buying foodgrains, often reflecting on the poor status of a farmer. Maize was plentiful in the *lekh* and so was *gur* (raw sugar). In fact Baitadi had a surplus for export through Jhulaghat to the neighbouring Indian district of Almorah. During World

[1]In Nepal, population has grown at the rate of 2.07 per cent, between 1961 to 1971, while in the hills and mountains by about 17 per cent. Meanwhile, the size of farm has shrunk to 0.65 acres per family.

[2]Former forest land, often marginal land. Farmers do not even bother to make terraces on these lands.

War II, there was as much demand for able-bodied men for the army as for the foodgrains from Baitadi. Baitadi now has a chronic deficit in food and much of this deficit is being met through imports from Kanchanpur, a *tarai* district in the far-western tip of Nepal. Most people live below the subsistence level.

The biggest problem confronting the inhabitants of Baitadi is the supply of fodder and fuelwood. Traditionally some of the villages on the Nepal–India border have depended upon the Indian side for forage, and it is not unusual for women to spend the entire day in the journey, which involves going up and down steep trails. In fact the need for forage is so pressing that occasionally women clamber down to the boundary bank of Mahakali River to fetch grass.

A change in this 'symbiotic' relationship between people on the two sides of the border is discernible, and there are restrictions on the free movement of people, which the border people find difficult to adjust to.

Nearly three decades ago, there were still good forests of oak and rhododendron in Baitadi and wildlife was plentiful. Pheasant, partridge, thrush, *chukor* partridge and *dhukur* (dove) could be had readily. In the author's house there was always *bharal sukti*[3] and fresh meat of barking deer. But wildlife has become rare: and the scarcity of fuelwood and fodder is further aggravated by the demand from dozens of new offices, and army and police establishments which can pay a high price for fuelwood. Few people plant trees such as *chir* pine and *tooni* (*Cedrela toona*) on waste land, or for that matter even such wild plants as *dhungio* (*Woodfordia fructicosa*) and *raindail* (*Rhus parviflora*). Only some affluent families maintain a piece of land exclusively for growing forage.

As in the past, a sizeable number of people from this district continue to become soldiers of fortune; they enlist mainly in the Kumaun regiment in India, as they are not welcome in the regular Gorkha regiments. These soldiers come from distinct castes—*thakuri*s and *chettri*s mostly: others go to Bombay to serve as watchmen. A large number of people from Baitadi, as well as from the adjoining districts of Dharchula and Dandeldhura, have moved down to Mahendranagar, the new township in the *tarai* district of Kaachanpur. Initially, they become squatters on *eilani* land, an acre or two of partially-cleared forest land, which after a couple of years becomes permanent farmland. Some bought a couple of *bigha*s of marginal land which their owners sold cheaply when the Land Reform Act was introduced in 1964. Some moved lock, stock and barrel and after living under near famine conditions in the hills for decades, found the *tarai* like heaven. They could now have two square meals a day—a luxury which they had never enjoyed before. Government service is another source of employment and during the Rana regime, petty officers exploited poor peasants by conniving

[3]Meat dried slowly over fire; venison and fish is also turned into *sukti*.

with the local *thalu*s[4], who had their own patrons in the government offices. Thus, hunger and exploitation compelled many families to move out from their villages and go to India or even to Burma where they worked in the silver and lead mines. A sizable population emigrates to India in search of seasonal employment as members of road gangs or as porters in hill-resorts such as Naini Tal, Mussoori and Simla. The new regulations about movement in the border districts of India may deprive the people from far-western districts of an opportunity to earn some money during the winter. Official circles in Kathmandu are aware of the situation, and development projects are now being sanctioned to provide employment to these people.

When the author was last in Baitadi, he enquired of a group of people from Purchaundi, a backward area, suffering from chronic food shortage, as to their purpose in coming to Jhulaghat. They replied that they had come to buy maize which had arrived at Jhulaghat under the American Aid pro-gramme. Unfortunately it was flint maize, and could not be milled by the local *ghatta*s (water mills). Some had come to earn enough money to pay off debts which they had incurred in buying girls for marriage.

Gobadi didi, the *pujarin* of the temple at Urugao, was both feared and res-pected, though a *deoki*[5]. She was specially invited to every house whenever a religious ceremony took place, and had lived a life, if not of splendour, at least of affluence. But now a frail figure well past seventy, she was literally on her death bed. In a way she symbolized the village. Apart from the fact that she had to face economic hardship, what troubled her most was the thought that there might not be enough hands around to carry her corpse to Sera, the cremation ground, a good eight kilometres away at the junction of Mahakali and Chamaliya. At Pallachaundly, time seemed to have frozen; the village presented a picture of decay and deprivation.

SOCIAL CHANGE

In the past decade or so, the border trade has undergone ups and downs. Besides *ghee,* which is the principal commodity for export, and wheat and forest products, railway sleepers, medicinal plants, bamboo products, etc., there is very little to sell. All the daily necessities, such as salt, *gur,* sugar, tea, kerosene, textiles, shoes, thread, batteries and ready-made garments are imported through Jhulaghat. The *sauka*s (Bhotiyas) still bring salt but it is no longer economical to buy it from them. The *patuwa* (thread-man) still

---

[4]Local elite who handle local litigation and settle problems involving land, intercaste marriages, community service, etc .

[5]A female worshipper, who is 'presented' to the deity for looking after the needs of the temple, such as fetching water, cleaning, etc. She remains unmarried, but engages in prostitution freely.

brings braids, *bindis,* bangles, etc. during the month of Bhadra (July/August) at about the time Gora[6] is celebrated.

Though many of the traditional rituals, festivals and customs are still followed in the hills of Nepal, the younger generation considers these little else than social occasions. The traditional social ties are fast disappearing and the dependency of the client class on the *thalus,* small or big, is almost gone. In this process, the hardest hit are the *haliyas* (plough-men, usually untouchable) and the landless peasants, whose patrons have either moved to Kathmandu or to some other part of the *tarai.* It is this collapse of the old economic order which is at the root of the widespread poverty in the hilly regions of Nepal. In adjusting to the changing times, however, the untouchables have fared better than the 'upper caste' people, whose options to engage themselves in gainful employment or some sort of enterprise, are limited. Taking the example of Pallachaundly again, inhabited as it is by the 'upper' and 'lower' castes living a life of interdependence, the *upadhaya* Brahmins exerted a feudal power over the latter, reminiscent of the middle ages. A *haliya,* for example, not only ploughed the land but served as a porter, woodcutter or as a sort of handyman. He was often bought from some other person and hence was not free to engage in independent jobs. Recently however, the *haliya* seems to have achieved some freedom. To cite one specific example, our own *haliya* refused to bring fuelwood, in spite of the fact that our caretaker wanted him to bring a full load. Before 1951, except for the top administrative officers who were deputed from 'Nepal', i.e. Kathmandu, the rest of the positions in the civil, police and militia were occupied by Brahmins and Thakuris. There was much feuding between these two groups, who competed for influence at different levels. These *thalus* were in effect the political agents of the erstwhile Ranas, who often acted also in liaison with the British Indian authorities across the border. They had bonded labour as well, though this was officially abolished during the rule of Prime Minister Chandra Shumsher Jung Bahadur Rana in the thirties. Forced labour was the rule rather than the exception in those days, but such customs by and large have disappeared. One can now discern certain egalitarian trends, which are a reflection of the growing literacy (19 per cent according to the latest estimates), from almost zero at the beginning of 1950, when the despotic regime was shaken off. District headquarters, many now connected with STOL airstrips, are becoming more and more urbanized; they have modern amenities such as roads, schools, electricity and piped water. If, however, the tempo of developmental activities has quickened, in great part due to King Birendra's periodic visits, the people's aspirations for a better life are also proportionately increasing. It is also equally true that the great majority of the people in the hilly region live below the poverty line,

---

[6]A festival for women marked by singing and dancing around a basket containing fresh cereal plants.

a thing which forces them to migrate to India every year.

The decline of cottage industry in the hills and mountains has contributed towards the deteriorating economic conditions in these parts. Goldsmiths, blacksmiths, shoemakers, carpenters, etc., are either without work or there is so little that they have to turn to other professions. They have also migrated to the *tarai*—in the Resettlement Projects, most of the applicants are from the professional groups.

There has been a similar decline of traditional social, religious and cultural institutions in the hills of Nepal. Old skills in wood-craft, bronze, stone carving, weaving, printing, painting, etc. are being lost. Even classical music, dance, and drama have been replaced by the *filmi* music and dance. A new cultural nexus is developing between the hill people and the urban population which is symbolic of the new social situation. The trends towards modernization are discernible throughout the hills as is evidenced by the widespread use of synthetic materials, transistors, watches, and acquisition of a new idiom. Two decades ago, in the far western district of Nepal, Nepali caps were rarely worn by people, but now these have become popular. The national dress along with the new textbooks which have now been uniformly introduced under the National Education System Plan, are part of a conscious effort to reinforce nationalism in the educational institutions of the country. It is, however, a paradox that while tourists have been waiting eagerly to visit the trans-Himalayan district of Manang *bhot* which lies on the north face of the Annapurna range, the Manangis[7] themselves display a material prosperity comparable to those living in the big metropolises, such as Kathmandu, Bangkok, Hongkong or Singapore, where they traded in medicinal plants. In the past when trade between Tibet and Nepal flourished, the inhabitants of Nyesyangba, an area within Manang district, traded through Mustang, or Nar Phu or Ghyasumdo, and on the basis of a *lalmohar*[8] obtained around A.D. 1789 they acquired special privileges in matters of trade with overseas countries, which still continue. Before 1962, these Manangis travelled on Indian passports and their trade was confined to India and Burma. But now having a full right to a Nepalese passport, they roam throughout the south Asian countries and in exchange for yak tails, skins, herbs and musk, they import manufactured goods—watches, nylons, transistors, etc., for which there is a ready market in Kathmandu. Of course, these luxury goods are smuggled across to India also. As this sort of trade is quite profitable, even those who had no right to a passport have now claimed and secured these documents, and agriculture and animal husbandry, the traditional way of life, have been relegated to a position of secondary importance.

---

[7]Inhabitants of Manang district, whose total population, according to the 1971 Census, does not exceed 7436. The inhabitants are Gurung, and Bhotiya or Tibetans who immigrated less than 80 years ago.

[8]A royal seal authorizing a person to a land or office.

Development of communications, particularly roads, is bringing the vast hinterland districts of Nepal into contact with the Indian culture and economy, resulting in changes in the life styles of the people. Christoph von Fürer Haimendorf (1973), has illustrated this point by saying that the loss of Tibetan trade has been offset by the improvement in communications linking the middle ranges of Nepal with India. India now provides cheap salt, which threatens the traditional salt/grain barter trade which used to provide the Bhotiyas with a large part of their income.

HILL MIGRATION

Due to shortage of food and underemployment of farmers during winter, migratory movements take place to the *tarai* districts or to the warmer valleys where winter pasturage is available. Some communities, such as the Sherpas or the *thakalis* engage in this seasonal migration, as do the hill people, such as the *dotiyals* from Dandeldhura and Doti districts. A majority of these are what have been described as reversible migrants, or people who shuttle between their village of origin and the lower areas, either seasonally or for specific purposes. In all such cases, social ties are maintained. Another type is the 'Lahure migrants', who meet their necessities through seasonal employment as agricultural labourers, skilled artisans, and long-term migrants —*lahures, durvans* and *chowkidars*.

Extended Farms

Some of these farm owners buy land in regions different from that of their village of origin; for example, people of Baitadi district now consider it a matter of prestige to buy land in the *tarai*, in Mahendranagar or anywhere in Kanchanpur district. But they still maintain links with their original home for participating in festivals, religious ceremonies, marriages, etc. As this tenuous link has no economic base, perhaps it will cease after some time.

A change in the life style of people who migrate from the hills to the plains is noticeable. 'All change to cotton mill-made cloth and men reported giving up their hill style *bhoto* (waist coat) and *kachhar* (short towel-like dhoti) and adopting the more characteristic shirt (often short-sleeved), shorts, *dhoti* or trousers of the plains (Elder, 1976, p. 96).

Ethnic identities are maintained even after having been in the *tarai* for long periods and most resettlers marry their daughters at the age of 15–16 and within their own community. Another significant thing which was found in one of the resettlement villages was the high per capita wealth of the Brahmins, followed by the Chettris and Magars. The *kami* (blacksmith) had considerably less per capita wealth than the other three groups and there was a marked difference in literacy levels within the different ethnic groups as is indicated by Table 15.5.

TABLE 15.5

Ethnic Difference as Reflected in the Level of Education in a Resettlement Village

Level of education	Ethnic Group			
	Brahmin	Chettri	Magar	Kami
No school, no literacy	32%	64%	54%	81%
No school, some literacy	30%	28%	38%	15%
Some school	38%	8%	8%	4%

Source: Joseph W. Elder et al., *Planned Resettlement in Nepal's Terai*, Institute of Nepal and Asiatic Studies.

Ethnic Diversity and Change

Throughout the central highlands, there is a multi-ethnic and polyglot population. The Chepangs, whom Hodgson called the Lords of Unredeemed Waste, had no other hunting implements than bows and arrows. However, they are now settled communities living in villages and are engaged in agricultural labour for the high-caste Hindus. In Pandrong village, where Newars, Nagarkoti, Magars, Kamis, Damias and Brahmins also live along with the Chepangs, the latter form an indigenous majority (462 families). The Tamangs, who have 55 families, still undertake share-cropping for the Brahmins. As the Chepangs fall into debt, the Brahmins take their land.

Upreti has studied the socio-cultural changes in the Limbuan, a vast area in the eastern hills primarily inhabited by the Limbu tribe. In Brahmin society, secularization of values has been progressing steadily. Onions, garlic and tomatoes, once considered taboo, are now freely eaten, and an overwhelming majority of Brahmin boys born after 1950 consume chicken and eggs. However, in deference to the sentiments of older people, such food is not cooked in the main kitchen. Liquor is now generally considered an acceptable drink among educated Brahmins. Brahmins still do not keep poultry. Another new custom is cutting the hair in the Western style and wearing a *sari*—instead of *phariya*—a printed cloth 9 to 13 metres in length.

Hindu culture has made steady inroads into the Limbu society itself. This is particularly related to what M.N. Srinivas calls 'sanskritization', which is the drawing into the Hindu fold of a non-Hindu tribe or immigrant. Changes in Limbuan society include the acceptance of caste hierarchy and Hindu religious and communal practices. Most Limbus have stopped sacrificing buffaloes at religious ceremonies; they have resorted to the practice of cremating their dead instead of burying them; and they offer cows' milk instead of blood and *raksi* to the local deities.

Hinduism has affected even such quasi-Tibetan tribes as the Byansis, which are closely related to other Tibeto-Burman speaking groups, referred to locally as the *sauka*s. They claim kinship with the Chettris and speak a Tibeto-Burman language. These Byansis living in the far-western part of Nepal have two major settlements at Tinker and Changru, with a total population of two thousand. They trade in salt, cloth and grain, and travel as far down in the *tarai* as Mahendranagar. Their religious practices are a combination of Tibetan Buddhism, hill animism and Hinduism. They claim the famous chronicler of the epic Mahabharata, Vyas Rishi, as their guru. Many are rich enough to send their children to schools in Pithoragarh, Askote or even to Kathmandu. Education has led to some Byansis working as overseers, school teachers, officers in the Nepalese army, employees in travel agencies in Kathmandu, and even as ministers. As border trade with Tibet has shrunk, these people have been forced to seek alternative sources of income.

Of the tribal groups inhabiting the forest, the *ban raja*s (King of forest) or the Kusundas, are a vanishing tribe. Reinhard (1977) has been able to locate a few families in central and western Nepal after much difficulty. Hodgson wrote in 1848: 'These people are as near to what is usually called the state of nature as anything in human shape can well be . . . . They were nomads, who lived in lean-tos and hunted with bows and arrows.' Reinhard found that none of the people he encountered 'knew much about the customs of their forefathers or spoke the *ban raja* language.' These remnants, who are descendants of mixed marriages, are now settled agriculturists. He found only one person who spoke the *kusunda* language and lived by hunting. Another vanishing tribe—the Rautes, claim that the Kusundas are headhunters, a thing stoutly denied by the latter.

In common with other Himalayan tribes, the *ban raja*s claim ancestry from the high caste (twice born) Thakuri, so much so that some have even changed their caste and call themselves Magar. According to Reinhard: 'Only a few hunt today and even they are dependent on odd jobs and working in fields for a part of their subsistence.' Rejection of nomadic life is due to the loss of the habitat—the forests, and hence they were forced to settle on farms. Now for all practical purposes, they have been absorbed in one or other of the organized tribes, such as the Magars.

Marpha is a small village in the central trans-Himalayan region. A significant new development is the new horticulture station where crops entirely new to the area, such as grapes, peaches, apricots, plums, almonds, cabbages, etc., have been grown successfully. Instead of *chang* (beer made from barley), fruit wine is becoming popular. The opening up of parts of Manang and Mustang to trekkers has created fresh economic opportunities for the local people. Trade will soon become secondary to tourism, as has happened with the Sherpas in Solokhumbu, a point which has been well emphasised by Christoph von Fürer-Haimendorf in his *Himalayan Traders*: 'Only twenty years have elapsed since I first visited Khumbu, but so great has been the change that it seems doubtful whether the traditional economic and social order which I then observed could be reconstructed by a study of the Sherpas as they are today. The complicated system of barter trade with Tibet, for instance, will soon vanish from the memory of the Sherpas now more and more drawn into a monetary economy subject to the inflationary trends characteristic in Nepal as well as India. Modern developments, such as shortage of agricultural labour, become explicable when seen against the background of the changes in the earning potential of landless men within the past sixteen years, and the contraction of the Sherpa's pastoral activities finds its explanation in the diminished opportunities for the cattle trade with Tibet which in 1957 was still flourishing.'

Be that as it may in that part of the country, in the border areas of Dolpo, trade has been resumed to a greater extent, largely because of the requirements of the Chinese administration in Tibet, especially for rice and sugar.

The highest settlement in Dolpo is at Charka (4350 m). The main economic activities are agriculture, cattle breeding and trade. Yak breeding is another important economic activity which has compensated to some extent for the loss of trade with Tibet. The houses are adapted to the local climate which is characterized by meagre rainfall and a high degree of insolation. The houses are made of rough uncut stone and mud mortar and have few windows. The roofs are flat and very little wood is used because it is so scarce. Only 'living rooms of especially wealthy people have a wooden flooring. In comparison to other areas of the Inner Nepal Himalaya, the interior furnishing with wooden cupboards and shelves is poor' (Kleinert, 1977). Around Dolpo there are abandoned fields and numbers of ruined houses which indicate that agriculture and the population are decreasing. Perhaps there could be no better proof than this of the fragility of the mountain ecosystem and that any sudden increase in local population such as through the demands of mass tourism is sufficient to tilt the balance against nature.

Langtang village, situated at an altitude of about 4000 m, is primarily inhabited by Lama Tamangs. In May 1974, a group of Khampas at Ghora Tabela had been encroaching upon a forest of pure hemlock. They were refugees who had been allowed to settle there but had become troublesome to the local residents. It was not unusual for a Khampa to abduct a married woman. The local inhabitants were also worried about the setting up of the National Park, for this meant loss of freedom to graze cattle. An old woman who seemed to be the spokesman of the villagers asked the Minister who was investigating the Khampa problem, 'What is it that the white men find so attractive here that we are being driven away from our own place?' Obviously, the people were agitated and also misinformed. Since then, however, the Khampas have been resettled elsewhere and there has been no major problem between the villagers and the Park authorities.

Sherpas of Rolwaling

Secherer (1977) in discussing the life pattern of the Sherpas has observed that the potato was introduced into this area nearly one hundred years ago. It is the major crop, and is bartered for lowland grains, and yak hybrids are sold for cash. All other items of consumption, including salt, teak, chillies, sheep fat and manufactured goods have to be imported. Secherer asserts that 'certainly, their demographic expansion, present-day food surpluses and general *joie de vivre* indicate a highly successful adaptation.'

The Sherpas have followed an efficient system of nature conservation; the yaks are never grazed for more than five years at one spot, and the herdsmen move continuously upwards until the cold weather forces them to make the return journey. In these high-altitude pastures which extend to above 4800 m, temporary shelters of stone and wooden planks have been erected. Their fear of being deprived of these traditional grazing grounds is aggravated because their other traditional grazing grounds across the border into Tibet are now inaccessible to them.

Another group of Sherpas live in Helambu, on the southern face of the Gosainkund range. They are different from those living in Solokhumbu. For one thing, they are not engaged as high-altitude porters and they are mainly agriculturists. Recently they have started fruit farming, although the quality of the fruit is not as good as in Kashmir or Kumaun. In 1975, according to Manandhar and Veerland (1976), a total of 48 tonnes of apples were produced, of which 18 per cent were sold to tourists.

Tourism

There has been a phenomenal increase in the tourist trade. While only 12,000 visitors came to Nepal in 1966, in 1975 there were 92,440, apart from 17,881 from India. In 1975, tourism contributed US $9,692,000 in foreign exchange, which was an increase of 20 per cent over the preceding year. Nearly a fourth of the visitors to Nepal are trekkers who visit Namche Bazaar, Helambu, Pokhara, Jomosom, including Annapurna Base Camp and the Langtang–Gosainkund area. Tourism, besides bringing employment to the dwellers in the high altitudes has also had profound sociological effects. Nepal's tourist spots are in the timberline zone, which is characterized by rough terrain and harsh climate. Vegetation is very sparse and whatever crop grows takes long to mature. Despite the increase of tourism there has been no increase in the production of foodgrains, poultry, dairy products and fuelwood. Certain areas have borne the brunt of the effects of mass tourism; in 1970, more than twenty expeditions were mounted in different parts of the country and some of the peaks were booked as far in advance as 1980! Large expeditions mount their operations with the same zeal as for a war operation—thousands of porters, helicopters and ski-planes take part in the assault. Trekkers to the region of the Everest Base Camp now cannot expect to find milk, poultry or meat on the way. They mostly live on potatoes, *dal* and rice which has to be brought from the plains. Fuel is in short supply and yak dung, instead of being used to manure the fields, is burnt. Besides, during the peak tourist season, local inhabitants have little or no time to attend to their farms or socio-cultural activities, and commercial attitudes have led to a decline in moral values. Mass tourism in the mountainous regions of Nepal 'may result in a negative approach to the development of tourism. Tourism is not only the goose that lays golden eggs but it also fouls its own nest' (Shreshtha, 1976, pp. 85–95). Food, firewood and other articles of daily use are no longer within the reach of the local people. The trekking routes are littered with all sorts of garbage—such as plastic containers, toilet paper, beer cans. The Lamosangu–Everest route, which lured more than four thousand trekkers in 1976 has become so polluted that drinking water is not considered safe, and the breath-taking scenery and solitude of these parts is being despoiled.

Tourism undoubtedly has a positive aspect as well. Inn-keeping has become a very lucrative business; the *bhattis* (inns) of the *thakalis* are a picture of neatness. A Sherpa has used an old discarded aircraft—a Dakota—as a restaurant in Lukla. Also, prospects for popularization of the local handicrafts have brightened. The establishment of National Parks and Wildlife Reserves in the Himalayan region has added a new dimension to its development. Undeniably, the future of tourism is linked with protection of the environment.

PROSPECTS

Nepal's development strategy is to put special emphasis on programmes which are aimed at removing regional imbalances between the hills and the *tarai* on the one hand, and between the various hilly regions on the other. The country has been divided into four development zones, each with a centre and growth corridor along a north–south axis. Depending upon the geographical situation, livestock in the mountainous region, horticulture in the hilly region and cereals in the plains (*tarai*) are being developed. Furthermore, through the Integrated Hill Development Project and Special Area Development Programme, special developmental programmes are being implemented in the hilly and mountainous regions of the country. Special attention is being given to village-level programmes in developing local weaning-foods, on nutrition and child care, education, food-preparation demonstrations, vegetable and fruit cultivation and school gardens and orchards. These activities in which the assistance of UNICEF has been secured, are part of an integrated hill-development programme designed particularly to improve the lot of rural women.

G. Toffin calls attention to the unprecedented degradation of the mountain environment in the last thirty years. Particularly noteworthy is the increasing numbers who have settled away from home; in 1951, 3.5 per cent of the population left their homes, and by 1971 the rate had roughly doubled. It is estimated that about 50 per cent go abroad—to India, Bhutan or Sikkim (Toffin, 1976, pp. 31–2). In spite of the large number of people serving in the Indian or British armies, a *lahure* after returning from the army readjusts himself to the village community life: as Toffin remarks, 'He begins to wear his traditional clothes again, keeping only one shirt or a pair of socks as a souvenir of his former status. The hygienic habits learnt in the army are forgotten. He soon goes back to the fields with his swing plough and basket on his back. Gradually, the village takes over.' Those returning from Bombay for a while wear fine Finlay *dhoti*s and round felt caps. They bring back such status symbols as radios or gramophones, which they soon sell to the local landlord. Their material prosperity does not last more than a couple of years and then they return once again to Bombay to serve as *durvan*s or watchmen.

The radio plays a key role, especially in popularizing Nepali music and songs, but the only constraints are that few people possess wireless sets and the programmes of Radio Nepal are not clearly heard in all parts of the country. On the other hand, the powerful Indian radio can be heard everywhere. So in a remote part such as Achham, one can hear Indian film music on a transistor, which has become a status symbol in the rural areas now.

In an attempt to bring the Tibetan-speaking community into the mainstream of national life, every year groups of lamas from border areas are brought to Kathmandu under the *Desh Darshan* programme. King Birendra spends a good deal of time outside the capital visiting remote parts of the country. In order to decentralize administration, regional centres have been established at Surkhet, Pokhara and Dhankuta in the far-western, western, and far-eastern development zones. The *Mahendra Rajmarg,* a six-hundred kilometre long highway extending from one end of the country to the other—from Mechi to Mahakali in the *tarai*; and north–south roads, such as the Pokhara–Sunauli road, and the proposed Pokhara–Surkhet road, to be built with Chinese assistance, are all measures which will strengthen developmental infra-structure in the hilly parts of Nepal. It is hoped that with the construction of the Dolalgha–Dhankuta road—a 322 km stretch running across the eastern midlands, Nepal will have an effective communication network between the hills and the *tarai* . This, besides integrating the economies of these parts, would also hasten the process of acculturation. The question of political leadership, which at the moment is vested with the 'Back to the Village National Campaign' has to be tested on the anvil of time. The *panchayat* polity has yet to mobilize the vast majority of the masses who live in the hills and mountainous regions. Therefore the future of the Nepalese midlands and the trans-Himalayan region can be viewed with only cautious optimism.

BIBLIOGRAPHY

BHATT, DIBYA DEO. 1977. *Natural History and Economic Botany of Nepal*. Calcutta: Orient Longman.

CAUGHLEY, GRAEME. 1969. *Wildlife and Recreation in the Trisuli Watershed and Other Areas in Nepal*. HMG/FAO/UNDP.

CRONIN, EDWARD W. D. 1977. 'The Problem of a Forest Remnant'. *Nepal Nature Conservation Society Annual 1977*.

'Development of Services Benefiting Children in Nepal'. Plan of Operation., 1975–80. UNICEF/HMG/UNESCO in Co-operation with HMG/Nepal.

DOBREMEZ, R. P. 1976. 'Exploitation and Aspects of Medicinal Plants in Eastern Nepal'. In *Mountain Environment and Development*. SATA (Swiss Association for Technical Assistance). Kathmandu.

ECKHOLM, ERIK P. 1976. *Losing Ground: Ecological Stress and World Food Prospects*. Ind. Printing. Delhi: Hindustan Publishing Corporation.

ELDER, JOSEPH et al. 1976. 'Planned Resettlement in Nepal's Terai'. Institute of Nepal and Asiatic Studies.

FLEMING, R. & R. FLEMING. 1977. 'Bird Movements in Nepal'. *Nepal Nature Conservation Society Annual 1977*.

FOREST RESOURCES SURVEY, Kathmandu, 1973. *Forest Statistics of the Hill Region*.

GURUNG, HARKA. 1974. 'The Population Aspect of Development'. *Population and Development*. CEDA.

GURUNG, NARESWAR JUNG. 1976. 'An Introduction to the Socio-Economic Structure of Manang District'. *Kailash* (4) 3.

FÜRER HAIMENDORF, CHRISTOPH VON. 1973. 'The Changing Fortunes of Nepal's High Altitude Dwellers'. *The Anthropology of Nepal*.

IBRD Report. 1973. Quoted by R. Hoegger in 'Mountain and Environment Development'. IAST & SATA. Kathmandu.

INSTITUTE of NEPAL and ASIATIC STUDIES (INAS). 1975. *The Effect of Out-Migration on a Hill Village in Far Western Nepal*. Tribhuvan University Press, Kirtipur.

KLEINERT, CHRISTIAN. 1977. 'Dolpo—the Highest Settlement Area in Western Nepal'. *Jour. Nep. Res. Centre 1 (Humanities)*: 11–24.

MANANDHAR, D. N. and C. C. VEERLAND. 1976. 'Fruit Production in Helambur and Impact of Fruit Cultivation upon the Economy and People of Helambur' (mimeo).

MANZARDO, ANDREW E., DILLI RAM DAYAL and NAVIN KUMAR RAI. 1976. 'The Byansi: An Ethnographic Note on a Trading Group in Far Western Nepal. *Nep. St.* 4(1): 1–21.

MAUCH, S. P. 1976. 'The Energy Situation in the Hills—Imperatives for Development Strategies'. In *Mountain Environment and Development*.

Nepal Tourism Master Plan. 1972. Ministry of Tourism, HMG, Kathmandu.

REINHARD, JOHANN. 1977. 'The Ban Rajas—A Vanishing Himalayan Tribe'. *Contr. Nep. St.* 4(1): 1–21.

ROBERTS, JAMES. 1977. 'Pheasant Conservation in Nepal'. *Nepal Nature Conservation Society Annual 1977*.

SAKYA, P. R. and DOBREMEZ, J. F. 1976. 'Provisional Notes on Vegetation of Region Biratnagar–Kanchenjunga'. Unesco Regional Meeting on Integrated Research and Training Needs in the Southern Asian Mountain Systems, in particular, Hindukush-Himalaya (mimeo).

SCHALLER, GEORGE B. 1973. *Jour. Bomb. Nat. Hist. Soc. 69*(3): 523–37.

SECHERER, JANICE. 1977. 'Man in the Himalayas: the Sherpas of Rolwaling'. Nepal Nature Conservation Society.

SHAKYA, KRISHNA MAN. 1975. 'Floods: What They Mean and How to Control Them'. *The Rising Nepal*. 9 August 1975.

SHARMA, C. K. 1977. 'River Systems of Nepal'. *Nepal Nature Conservation Society Annual 1977.*

SHRESTHA, K. K. 1975. 'Effect of Tourism on Mountain Environment in Nepal'. Draft Report No. 34. Unesco Regional Meeting on Integrated Ecological Research and Training Needs in the Southern Asian Mountain System.

STAINTON, J. D. A. 1972. *Forests of Nepal.* John Murray.

——. 1976. 'The Impact of Tourism on Mountain Development'. *Mountain Environment and Development* 85–95.

STERLING, CLAIRE. 1976. 'Nepal'. *The Atlantic.* October 1976.

TAUTSCHER, O. G. 'Surveys and Demonstration for the Management and Development of Trisuli Watershed'. UNDP/HMG.

TOFFIN, G. 1976. 'The Phenomena of Migration in a Himalayan Valley in Central Nepal'. *Mountain Environment and Development.*

Tribhuvan University, Institute of Science. 1976. 'The Energy Sector' (mimeo).

UNESCO. 'Impact of Human Activities on Mountain Ecosystems'. MAB Report No. 8.

UPADHAYA, R. M. 1976. 'Food and Fodder as Energy Resources in the Hills of Nepal'. Seminar on Development and Management of Energy Resources in the Hills. IAST/SATA.

UPRETI, BED PRAKASH. 1976. 'Limbuan Today: Process and Problems'. *Nep. St. 3*(2): 47–70.

WEBER, WILLIAM. 1975. *Newsletter.* Special Coronation Issue. Nepal Nature Conservation Society.

WEGGE, PETER. 1976. *Himalayan Shikar Reserves: Surveys and Management Proposals.* FAO/UNDP/HMG, Rome.

WEINER, MYRON. 1971. *The Political Demography of Nepal.* Seminar on Population and Development. CEDA.

CHAPTER 8

LADAKH: DEVELOPMENT WITHOUT DESTRUCTION

HELENA NORBERG-HODGE

Ladakh, or 'Little Tibet', is a harsh and mountainous desert of wild beauty, set among the jagged peaks of the western Himalaya. Lying north of the main Himalayan watershed, beyond the reach of the summer monsoon, it is an exceptionally cold and dry area, with an annual rainfall of less than 12 cm and temperatures dropping to as low as − 50°C. Its long winters are as severe as those of almost any inhabited region of the world.

Yet, despite these conditions, the Ladakhi people not only manage to survive, but are able to enjoy a life of greater prosperity than that of many other Himalayan peoples whose natural resources are more abundant. By the standards of the developed world, it is far from being a comfortable life—small fires of dried animal dung, for instance, make little impression against the extreme cold of winter—but everyone is at least able to cover his basic needs, and no one suffers the terrible poverty and misery that are found in so many parts of the Third World.

With the exception of the capital, Leh, a town of 10,000 inhabitants, which has always been predominantly a trading centre, the vast majority of the 100,000 Ladakhis live by subsistence farming in small village communities, working plots of between two and four acres. Until recently, this traditional way of life had seen little change over centuries; in the last fifteen years, however, Ladakh has become more and more exposed to outside influences, to the extent that its culture is now severely threatened.

Since 1962, when trouble with the Chinese first flared up, a large contingent of the Indian Army has been stationed in Ladakh, and convoys of trucks can be seen every day bringing in supplies to Leh and beyond. Their presence, without causing devastating effects, has gradually corroded the local culture. The money economy, which had previously operated only among a few traders within Leh, is now much more widespread, and an ever-increasing range of imported goods is available in Leh's bazaar. Slowly but noticeably, the offshoots of the consumer world are penetrating Ladakhi society: corruption is increasing; theft is no longer quite the rarity it used to be; disparities in the standard of living are gradually appearing. Moreover, young men are being lured away from their traditional work in the fields by attractive financial rewards, so that it is now the common practice for one son from each household to enlist in the army.

The army presence has not yet, however, endangered the foundations on which traditional Ladakhi society is built; in fact, in the remoter villages, even its superficial effects have been slight. It is two further influences that could, if not carefully controlled, prove to be more destructive: India's own development programmes, which, especially since the arrival of the army, have come to include Ladakh and, less directly though equally potently, tourism, which in three years has already become a major industry.

This chapter is not directly concerned with change that has already occurred in Ladakh. Instead, it looks at the traditional society that, outside Leh at least, still exists virtually intact, and discusses how that society can continue to survive in the face of the considerable pressures that are now being exerted on it.

The success of traditional Ladakhi life is the result of the most prudent and frugal use of the very limited natural resources. The almost total lack of precipitation means that irrigation is a major problem, and is in fact largely responsible for determining the location of each village. In the absence of pumps, the many rivers that run through Ladakh's deep, barren valleys cannot be used for irrigation purposes, and so the farmer has to rely instead on melt-water from the snow-capped mountains far above. This water is led down through a wonderfully ingenious series of channels (*yura*) up to five kilometres long, to small patchworks of fields which have been painstakingly constructed out of the very dust and rocks of the desert. Every last square patch of irrigated land is used, since the severe climate allows cultivation for only five or six months out of twelve. In this time, the Ladakhis have not only to meet their day-to-day needs, but also to provide enough food for themselves for the long winter months.

Altitude and topography determine the choice of crop but barley, which is roasted and ground to form the staple *ngamphe*, is by far the most important. *Ngamphe* does not require cooking, and is therefore particularly suitable for consumption in winter, when fuel is extremely scarce. Many types of fruit and vegetables are also grown, and are dried in the sun before being stored away for winter. Peas (*shanma*), of which there are four varieties, are the most common vegetable, and can be easily preserved; apples and apricots are the main fruits, the latter being a valuable source of Vitamin C.

Although the land is used to the utmost, no attempt is made to work it beyond its capacity, and the Ladakhis are scrupulously careful to ensure that what is taken from the land is returned. Human nightsoil as well as animal dung is used as fertilizer so that, even without fallowing or crop rotation, good yields are rendered possible year after year. The trees that are grown—the poplar and the willow—are reforested, and not lopped indiscriminately, as elsewhere in the Himalaya.

The Ladakhis' frugality is remarkable. Nothing is wasted that could possibly serve a useful purpose. The same barley that has been fermented to make *chang* is then dried and roasted for eating; the water in which dirty plates are washed is specially kept to supplement animal fodder; clothes that have been worn until they are literally in tatters are used to dam up irrigation canals; even the crushed remains of apricot kernels, from which oil has already been extracted, are used—either to form a base on which wool spindles are rotated, or added to animal fodder to cure digestive disorders. Every form of tree and bush is also put to good use: poplar and willow to make furniture with and in the construction of houses, the finer apricot and walnut woods for musical instruments, wild juniper bushes for use in religious ceremonies, thorn bushes for building fences, and other bushes for baskets, fuel and medicine.

Next to agriculture, animal husbandry is the most important activity of the Ladakhi farmer, since dairy products are an essential part of his diet, and wool is needed to make clothes and blankets. As every piece of irrigated land around the villages is cultivated, and unirrigated land is effectively desert, the shepherds take their animals to the high mountain pastures, where there is more land available for grazing than at lower altitudes. This often requires a trek of several days up difficult gorges, after which the shepherds spend the summer in simple stone huts, tending sheep, goats and yaks at elevations of 4000 m and above. The shepherds send milk products, dung and wood down to the villages on the backs of donkeys and mules. These animals then return with supplies of food.

This combination of sheer hard work and an economic use of natural resources is the immediate reason behind Ladakh's relative prosperity, but the real roots of this prosperity are to be found in the traditional social structure. A look at a number of specific social institutions will show how fundamental this structure is to the success of Ladakh today.

One of the great strengths of Ladakhi society is that over the centuries a tight control has been kept on population. At least partly responsible for this has been the practice of fraternal polyandry which, although technically banned in 1941, is still to be found in the remoter villages. By this system, a woman marries into a household and becomes the wife of all the sons; population growth is thus strictly regulated since, although the numbers of men and women are equal, relatively few women marry and bear children. The practice of polyandry is now gradually dying out since not only is it illegal but it also deprives many women of what is now being seen as their right to marry and raise a family. Its role in controlling population, however, has been enormous, and it is a matter of some concern that, as a result of its decline, a noticeable, though so far not alarming, increase in population has already occurred.

Co-operative labour (*bes*) is an important feature of traditional life and encourages the full use of human resources. Although each household farms its own land, and each member of the household does his own daily work on that land, the community also works as a unit. Farm instruments and draught animals, for instance, are often shared, and individuals will take turns in watching over the animals. Irrigation, too, a matter of the utmost importance, is a communal activity. At harvest time the whole village will join in and each man lend assistance to the other. Indeed, such is the spirit of co-operation that the times of harvesting even of different villages will sometimes be deliberately staggered, to enable each village to help the other in its work.

Mutual support is in fact seen in all areas of Ladakhi life. The *paspun* is a good example. This is a brotherhood of between four and ten households who provide special assistance at the time of the three great events of life—birth, marriage and death. At births and marriages, for instance, they provide and prepare food for the celebrations; after a death, they arrange the funeral and relieve the close relations of the unpleasant task of handling the body. Each man offers his assistance spontaneously and voluntarily, and can be sure that, when he in turn is in need, the same help will be forthcoming.

On a broader scale, individuals in each village are appointed to a variety of posts which involve the well-being of the community as a whole. The *churpon*, for example, supervises the irrigation process, dividing the water out equally among the fields; the *lorapa* rounds up stray animals in order to prevent crop damage. These posts are normally held for a period of one year, but there is considerable flexibility in the system, and terms of appointment can be lengthened or shortened as required.

Ladakh's social traditions, in other words, work in the people's favour. Population is kept at a level at which it is possible to satisfy everyone's basic needs, and the communal nature of the social structure allows that possibility to be realized. A comparison with the lives of the neighbouring Baltis, whose resources are identical to those of Ladakh but whose standard of living is appreciably lower, underlines the significance of this point.

It is not strictly true to say that every farming household in Ladakh is self-sufficient; special skills—weaving and tailoring, for instance—are practised by only a few. But the further needs that the individual does not satisfy for himself will almost always be met from within the village in which he lives, so that reliance rarely has to be placed on anyone outside the immediate community of which he is a part. Salt and tea have traditionally been almost the only imported commodities. It is true that, as a result of the army's presence, more and more products have become available—kerosene for cooking purposes, and sugar, are two of the most popular—and bartering is also carried on with the nomads, but this trade merely provides what are considered luxuries, and is in no way depended upon for survival.

Besides being relatively prosperous in material terms, Ladakhis are an exceptionally good-natured, contented people . Their faces are lightened by almost constant smiles and even the most routine work is carried out to the accompaniment of song. Visitors are always astonished to find that a people who lead such a hard and physically uncomfortable life should appear to derive so much satisfaction from it. Moreover, those Ladakhis who have experienced the greater comfort of life elsewhere have almost invariably chosen to return to Ladakh. There is still virtually no crime, and disputes of any kind are rare. Even critical operations such as the sharing of water for irrigation, where friction between one farmer and the next might be expected, are conducted in a spirit of friendliness and mutual trust.

They are a healthy people too. Medical services are provided by local doctors (*amchi*), who will usually have received many years' training in Tibetan medicine. There is little serious illness—though tuberculosis and cataracts are both relatively common—and everyday ailments are treated with herbs. Alternative treatment of a less physical nature can be obtained from a *laba*, who achieves inspiration through self-hypnosis. His principal role, however, is that of an oracle, and he is consulted about the auspices of a wide range of matters of day-to-day importance.

The monasteries which form such a conspicuous part of the landscape, traditionally held great power in Ladakh. Today they are no longer the centres of administration, and the previous links with Tibet have been severed since the closure of the border in 1959. Nevertheless, they continue to serve the spiritual needs of the people and to hold a valuable place in the community. It is important to remember that Tibetan Buddhism is not the only religion found in Ladakh; substantial communities of Muslims exist in certain areas and there are also some Christians. A significant feature of Ladakhi society is that all these different groups live in complete harmony, each respecting the beliefs of the other.

Traditional Ladakhi life, even as it is today, is strikingly similar to Gandhi's Utopia, the ideal he believed had existed only in the agrarian villages of ancient India. Gandhi envisaged a republic of small village communities, each one self-sufficient. A man would do his 'bread-labour'—daily manual labour sufficient to provide for his own primary physical needs—and, at the same time, work as part of the community in order to satisfy the needs of the village as a whole. There would be a minimum of laws and other restraints, and decisions would be truly democratic, since communities would be small enough for each man to have a say in how his should be run. There would be no crime, nor discrimination in terms of caste, religion or sex. Disparity of wealth would not exist. Each man and woman would have an equally valuable role to play in society; no one would benefit at the expense of another.

Ladakhi society, which contains so many parallels with that ideal, is, as we have already seen, now being threatened. How, then, is it possible to prevent that threat from being realized ? How can Ladakh be developed without thereby being destroyed?

Conventional development theories have always stressed the importance of material growth, and have used purely economic yardsticks as measures of progress. A country's GNP, for instance, has been used as an indicator of its well-being. Bigger has always meant better. Giant organizations have sprung up throughout the industrialized world, making necessary giant cities and therefore still more giant organizations. Production has escalated as processes become ever more mechanized. Complex and increasingly rapid systems of communications have been established; promotion and sales networks spread worldwide.

But to what avail? The truth, which is forever becoming more apparent, is that economic growth in itself is not necessarily followed by an improvement in the conditions of life, in terms of the welfare and contentment of the people. Far from it. As the British economist, E.F. Schumacher, points out, the consequences of development based purely on the pursuit of material growth, are squalid urbanization, pollution, waste and dehumanization of work and community life. All these are the inevitable results of a system that thinks first of production and only then of Man himself.

Development in Ladakh should be rurally based and preserve, to the greatest possible extent, the Ladakhis' self-sufficiency. Each man should still be able to do his bread-labour and thus continue to provide the necessities of life for himself. This would not preclude the introduction of new skills and techniques which might help to increase his efficiency. Indeed, so long as it can be adapted to meet Ladakh's particular needs, modern technology could play an important part in its development; simple machinery, that could be made and maintained locally, could obviously provide considerable benefits. Cottage industries, which have already received the support of the Indian government, could be further encouraged. Operations, however, should always be on a small scale, allowing the individual to be personally involved in his work and giving him the satisfaction of seeing that the end product and the quality of that end product are the results of his own efforts. Fortunately, centralization of labour would be difficult in Ladakh since villages are scattered far and wide throughout the region.

Development should also be gradual. Any sudden change in working systems must be avoided even if it means, as it inevitably will, that material progress is thereby slowed down. A major contributory factor in the social decline that followed industrialization in the 'developed' countries of the world was that people were just not capable of keeping pace with the enormously rapid advances of scientific achievement. In a society such as Ladakh's, which has seen so little change over centuries, this is particularly important. Moreover, it is practicable since, in contrast with many other areas of India where extreme poverty dictates that material improvements be as rapid as possible, Ladakh can afford the greater time required to ensure that full consideration is given to the crucial matter of long-term social welfare.

All forms of development should be aimed at helping the Ladakhis to help themselves, so that their lives remain essentially in their own hands and not subject to the whims of other bodies beyond their control. Local resources, for instance, should be used wherever possible and local people trained to exercise any necessary new skills. Reliance on large-scale importing, of either goods or services, should be discouraged.

Development based on the lines set out above may be less attractive politically than the traditional alternatives; after all, new towns, factories and power plants can all be offered as immediate 'proof' of progress, while to pinpoint human contentment is a rather more difficult task. Nevertheless, there is some hope. In Ladakh, there is an exceptional opportunity to prove that Gandhi's teachings, which some would argue are intrinsically unsuited to the demands of modern living, can still be most effectively applied.

Finally, mention must be made of tourism since, of all the threats to the local culture, this may well be the most dangerous. Ladakh was opened to foreigners in 1974 and in that year only 200 tourists visited the region. In 1977, the figure was 15,000. If that figure is allowed to escalate still further, it will not be long before the economy is very significantly altered and the traditional culture debased. In Leh, in fact, that process is already under way. The government of Bhutan limits the number of tourists as well as the areas open to them, and requires a minimum daily expenditure. Such a policy in Ladakh, which has already been urged upon the Indian government, would lessen the negative impact of tourism and bring money into the region in a controlled way, so as to benefit not only a few entrepreneurs but the entire community.

Gandhi once said, in reference to development, 'The supreme consideration is Man.' If these words can be remembered and the happiness of the Ladakhi people thus maintained, not only Ladakh itself but the whole of India will reap huge rewards.

BIBLIOGRAPHY

BUSQUET G., DELACAMPAGNE, C. 1977. *Ladakh*. Paris: Buchet/Chastel.
CUNNINGHAM A. 1853. 'Ladakh, Physical, Statistical and Historical'. Reprint New Delhi: Sagar, 1970.
DATTA, C.L. 1973. *Ladakh and Western Himalayan Politics*. Delhi: Munshiram Manoharlal.
DREW, F. 1875. *The Jumoo and Kashmir Territories*. Reprint Delhi: Oriental Publishers, 1971.
DUNCAN, J.E. 1906. *A Summer Ride Through Western Tibet*. Glasgow: Collins.
FRANCKE, A.H. 1907. *A History of Western Tibet*. London: Gollancz, 1907.
———. 1914. *Antiquities of Indian Tibet*. Calcutta, 1914. Reprint New Delhi: S. Chand.
———. 1901. 'Ladakhi Pre-Buddhist Marriage Rituals'. *Indian Antiquary* vol. XXX.
HEBER, A. and K. *Himalayan Tibet and Ladakh*. Reprint Delhi, Ess-Ess Publications, 1976.
HEDIN, S. 1922. *TransHimalaya*. Leipzig: Brockhaus Verlag, 1922.

CHAPTER 9

LOST VISION:
ART IN THE HIMALAYAN REGION TODAY

B.N. GOSWAMY

For one used to looking at its art, the Himalayan areas in the seventies present a depressing contrast to the situation of not more than two hundred years ago when the entire region seemed to be alive with creativity, when it was possible to go to the tiniest principality in the hills and come upon works of art that were products of a chaste, integrated vision. There is none of that faith with which a world-view was once presented in its art, little of the warmth and the richness of human elements it once possessed. In the late seventies one is greeted by an arid spectacle of mechanical repetitiveness in the arts, a living on the past, as it were. And no one is troubled by this. It appears as if art does not really matter to anyone any more.

Perhaps some attempt at understanding the complex causes of this pheno-menon would not be pointless. The vast geographical extent of the areas in which Himalayan art once flourished, and the time-span which is covered by it, are such that it is possible to approach the subject here only in terms of very broad generalizations. And yet even these may be of some use: the many valuable studies of the art of the Himalaya by many very distinguished scholars, centre round specific areas and deal with 'their own' problems of art-history; possibly an overview, however quick, would be of value.

It would be legitimate to divide the art of the Himalaya into two broad but clear categories: the hieratic and the non-hieratic. In the former category would easily fall sculpture and painting which self-consciously served not only religious but clearly defined monastic ends. In the very high Himadri ranges, areas like Tibet, or in the upper Himalayan areas like Ladakh, Lahul, Spiti, Nepal and Sikkim, work of great intensity was once being done. Religion was putting art to its own use, but in a resolute and inspired manner. There were here, as in all hieratic art, attempts to portray a closed but self-consistent world which kept some kind of pace with developments within the faith, alike in the folds of Lamaistic Buddhism and of Hinduism. The icon is what is most typical of this art; whether it is a *Prajnaparamita* or a *Tara,* a *Bhairava* or a *Yamantaka,* the images have a sense of immediacy about them.

They represent with singular force what the artist saw within himself and what a devotee, monk or layman, wanted to look at. For centuries together, in sculpture or murals, miniatures or ritual objects, it was religion which served almost exclusively as subject matter.[1] There were some influences that came in from outside, impulses that were registered with varying degrees of understanding, such as the art of the Chinese or the Mongols, the styles of Western Asia or India, and there was a certain amount of interaction between different expressions. Artists travelled both into and out of the area. We find Pala artists going to Nepal, Nepalese artists going to Tibet, Kashmiri artists working at Alchi, monks on the move all the time. But what remained *dhruva,* unchanging, was the nature of the vision even as it was being expanded and deepened. The repertoire of the artist was occasionally enlarged; his skill was marginally affected; his technique underwent some change; but his world-view remained clear in its essence. The painter and the sculptor drew sustenance from given subjects with their given iconography; the patrons earned merit from commissioning works on the same themes; heaven was important, and so was hell. The demons were very real, the powers of good truly beneficent.

This was more or less the situation for something like a thousand years. One can obviously overstate the fact of continuity, or overstress the static nature of the work,[2] but at least two other points can be made: one, about the enormous quantity and the high level of work done in the region, and the other about the affinity between the art and the culture in which it was produced. Today, much of this is changed. In a sense, the quantity of work being turned out is still impressive: apart from the renovation of old murals in places like Likir, Tikste, and Hemis in Ladakh, the *thangka*s that refugee Tibetan artists produce today in Sikkim or in Dharmsala,[3] or the countless 'bronzes' being cast in workshops in Nepal or Delhi (and sold for the most part to tourists) are to be seen everywhere. And not all of it is inconsiderable in terms of quality; in fact, some of it has been taken somewhat seriously by collectors and museum curators alike. But what is really lacking in it is the sap, that vitality which coursed through the earlier work. For most of it is the product of an ossified vision; the repetitiveness palls; the artist no longer

[1]This would appear to be a general statement to which exceptions can be cited. But it does represent the truth about this art in substantial measure.

[2]In the Himalaya, as in Tibet, it is no longer feasible after about the eleventh century A.D. to assign a work of art to a period, school or artist on the basis merely of distinguishing their manners and styles. Nor did geographical vicinity determine the style because the monastic studies entirely depended on models sometimes procured from distant centres of Buddhist learning and copied for hundreds of years. Thus, Madan Jeet Singh in *Himalayan Art*, (Greenwich, 1968).

[3]There are, in fact, regular workshops where these *thangkas* are being produced and sometimes sold as antiques by dealers to unsuspecting buyers. It is also possible to pick up for petty sums wood-cuts based on old *mandala*s designs, copied exactly from very old works.

seems to reflect the urges of the society which in itself has changed, for he is at a considerable distance from it, certainly psychological, quite often physical. His *samadhi* has become *shithila* (slack), so to speak: art is no longer a matter of spirit, but of money. One has come a long way from the times in which the art of these regions, with all its hieratic insistence, was alive, and created areas of dazzling beauty.

In the non-hieratic category of art in the Himalaya, the issues are different, and it might be interesting to trace these developments that take a radically different course.

One could take, as most typical of this kind of art, painting in the Pahari region, for it is one of the most intensively researched of Indian art expressions.[4] Here, as an illustration of the many changes which came about in non-hieratic art, one is able to sense a perceptible and quicker degree of responsiveness to changes of various kinds; political, religious, social and economic. To the layman, mostly because he sees the glossy volumes that are reasonably priced and aim at a given public, most of this art might appear to be predominantly 'religious' in its subject. Much is omitted from this popular coffee-table book view: the portraits, the satirical themes, the references to historical events, the caricatures,[5] the recording of pilgrimages.[6] But even if, in terms of quantity, it is demonstrated that religious-poetic themes take a precedence, the approach of the artist remains, for the most part, non-hieratic. He was interested in producing intensely moving human images, mythological figures cast in every human situation. Inter-penetrated with illustrations of religious works, the *Ramayana,* the *Mahabharata,* the *Bhagavata Purana,* the *Markandeya Purana,* the *Sivamahimna Stotra,* are themes which lean upon works of poetic literature that were avidly read in the hills. The archetypal figures in these works may have been religious or mythological, Krishna and Radha, Rama or Siva, but the appeal really is that of the poetry that inheres in the subject. The great sets that we know of: the *Rasamanjari* of Bhanudatta, the *Gita Govinda,* the *Sat Sai* of Bihari, the *Rasikapriya* of Keshava Dasa, the *Sundara Sringara* of Sundara Dasa, the *Naisadhacarita* of Sriharsa, are all works in which legendary and religious figures occur repeatedly, but the main concern is a lyrical depiction of such themes and the archetypal figures are used to enrich the content, to evoke associations which broaden the impact of the verses, and conjure up

---

[4]Prominent scholars are W.G. Archer, Karl Khandalavala, and M.S. Randhawa. There are several others of course who have also taken serious notice of Pahari painting in recent years.

[5]Some of the most delightful of these caricatures make fun of Vaishnava saints and their activities. S.S. Gupta reproduced some of these in his 'Catalogue of Paintings in the Central Museum, Lahore', and in *Modern Review,* Calcutta, 1922.

[6]Several drawings showing pilgrimages or journeys are found in public and private collections. The artists attempt to show both the pleasure of these visits and the difficulties inherent in them.

images that are universal in their reach.[7]

There is a great deal more to Pahari painting than this, to be sure. But even if one confines oneself to a study of these religious and poetic works from the area, it is possible to assert that the character of the art in general is by no means static, unchanging in quality. The active period within which Pahari painting flourished is about 250 years. In this period, short as it is compared to the millennium or so of which we speak in relationship to the hieratic category of art in the Himalaya, there are not only numerous expressions of it; within each there are significant changes which are clearly visible, not merely subtle and imperceptible.

The changes occur for several reasons. The Pahari areas were, for one thing, not as isolated and cut off from the rest of the world as is sometimes believed. Impulses from the major centres of culture in the plains of northern India were received with surprising swiftness through different channels. Not only do we know of frequent incursions into the hills by adventurers, we also know of the hills serving as a refuge to princes and generals fleeing from pursuing armies. There were also more peaceful channels of communication between the hills and the plains, like for instance, the great Indian institution of pilgrimages. The regularity with which rulers of the hill states, their families, and the people in general, went out to the plains for earning merit by bathing at Hardwar or Kurukshetra, or securing *darshana* at Gaya or Jagannath Puri, was singular. A painting in the hand of the great Pahari painter Nainsukh, showing an idol at Allahabad,[8] a temple with the three Jagannath images in the tiny Pahari principality of Jasrota,[9] a charming drawing showing pilgrims bathing in the nearly dry bed of the river at Gaya,[10] are among the more obvious results of this travel. The large number of entries by Pahari visitors one finds in the records of the *panda*s (priests) at centres of pilgrimage in the plains indicates the extent of such traffic.[11] At the same time there were people from the plains visiting the great hill shrines of Jwala-mukhi or Kangra Bhavan. These must have acted as equally effective channels through which influences travelled. The long series of *patta*s (records) and letters that are still maintained by the *panda*s of these temples are a fair indication of this. It is only geographically that the Pahari areas were removed from the plains. The quiet flow of men and ideas went on even while there was bloodshed between rulers and armies marched up and down the Panjab.

[7]It is unnecessary to refer to this body of work in any detail, because it is much too well known. Almost always, it is a poetic simile or conceit that the painter draws attention to.

[8]This painting showing the idol of Beni Madho is in the Bharat Kala Bhavan at Banaras.

[9]See Karuna Goswamy, 'A Pahari Painting of Jagannath Temple', *Chhavi*, B.K.B. Golden Jubilee Vol.

[10]This very attractive large drawing was once in the collection of the art dealer, Mohan Lal Bharany of Amritsar.

[11]For more information on this, see B.N. Goswamy, 'The records kept by priests at centres of pilgrimage, etc.', *The Indian Social and Economic History Review* vol. III No. 2.

Yet another channel of communication between the hills and the plains was established during the period of Mughal sovereignty from the beginning of the seventeenth century till the middle of the eighteenth. This is when the rulers of the Pahari areas were frequently commissioned by their overlords, the Mughals, to take charge of campaigns in different parts of India. Man Singh of Guler serving in the Northwest Provinces,[12] Raja Basu of Nurpur going out to Rajasthan,[13] or Ajmer Chand of Bilaspur moving down to the Deccan under Imperial orders,[14] are typical cases. Thus when we find Mughal costumes appearing in the hills very soon after they had been adopted at Agra or Delhi or Lahore, or Mughal and Rajasthani architecture affecting building styles all over the hills or when a great work like Bihari's *Sat Sai* comes to the hills within a decade of its having been composed in the plains,[15] it is symptomatic of the situation. Understandably, the initial impact was felt in the outer hills, but it travelled fairly quickly, in ever-widening ripples, further into the interior. What was registered in the states of Nurpur or Bilaspur or Basohli often reached states which were contiguous very quickly, and became evident all over the hills the next day at an unprecedented speed. In terms of stylistic development, in the field of Pahari painting it may be possible to see three major waves of influence from the plains.

The appearance of Pahari painting in the first half of the seventeenth century coincides roughly with the coming of what has been called 'ardent *Vaishnavism*' to the hills. In Kulu, Mandi, Kangra, Nurpur, Chamba, Jammu, missionary activity through Vaishnava saints spread quickly, bringing inspiring new subject matter, and inevitably affecting the modes of thought and expression in the entire area. The impact on painting was deep and pervasive.

The next major development in style in the Pahari area came at the beginning of the eighteenth century, apparently as a result of Mughal influence. The Mughal presence in the hills had been felt earlier but it is at this point of time that the painters of the Pahari area seem to have responded to Mughal works most warmly. This may have been due to Mughal paintings in large numbers reaching the hills, and not due to refugee artists from the Mughal court going into the hills, as is sometimes believed.[16] Whatever the circumstances, the shift towards a fluent, naturalistic style from the

---

[12]The *Diliparanjani,* manuscript history of Guler, makes prominent mention of this commission to Raja Man Singh.

[13]The Emperor Jahangir, in his *Tuzuk* (Indian Reprint, 1968) vol. I, p. 200, makes mention of Raja Basu campaigning against the Rana of Udaipur.

[14]M.A. Macauliffe, *The Sikh Religion* (Indian Reprint, S. Chand, 1963) vol. V, p. 165, draws attention to this visit of the Raja to the Deccan

[15]This copy of the *Sat Sai* was at one time in the collection of Sh. Bhu Dev Shastri of Sujanpur Tira in the Kangra Hills.

[16]The date often given for this 'migration' is 1739, the year of Nadir Shah's sack of Delhi. But I have tried to show in my 'Pahari Painting: The Family as the Basis of style", *Marg,* Sept. 1968, that this assumption is questionable.

earlier, highly mannered and intense style linked with the name of *Basohli* can be assigned to the period that we are speaking of. The change is by no means sudden, the artist explores various possibilities, makes adjustments, engages in experiment, and takes from Mughal painting what agrees with his own temperament to incorporate it into his own style.[17] Change is effected most intelligently, brought about apparently as a result of a warm understanding between patron and painter. The process can be seen in the work of gifted artists like Manak of Guler, or his younger brother Nain-sukh who worked for Balwant Singh of Jasrota. By the middle of the eighteenth century, the style had become significantly different from what it was in the seventeenth century. Responses to the new situation had been slow but extremely cultivated.

The next major change in the Pahari area came about in the nineteenth century, first as a result of the political and economic decline of the hill states and the corresponding rise of the Sikh kingdom in the Panjab, and then, from the middle of the century onwards, due to European impact when the area passed under British supremacy. The first part of this change had more than merely political consequences. The artist had now the clear option either to look for patrons other than rulers in the Pahari area, or to move in search of patronage towards the new centres where money and power now resided. A series of remarkable documents that have fortunately survived[18] tell us about the decision made by at least one major family of Pahari painters. They decided to move to serve the new masters at Lahore. Among the people they were attached to were not only the Maharaja himself but nobles closely related to the Maharaja, like the Sandhanwalia Sardars. This group of documents clearly shows painters moving to the plains and becoming attached to new patrons, while retaining their little *jagirs* in the hills where their small houses and plots of land were. These *muafis* had sustained them and their families over generations, and these they were allowed to enjoy through royal orders given to the local Sikh Governors of the hill areas.

The political situation in Lahore changed once again, however, with the fall of the Sikh Kingdom. The new masters, the British, were not interested in the art or themes of the Pahari painters, and the painters had to fend for themselves yet again. They turned to royalty for patronage once more, because royal houses had survived in some of the cis-sutlej states like Patiala, Kapurthala, Nabha and Jind. We have remarkable evidence of this kind of migration in the movements of an artist like Devi Ditta of Basohli who moved first from the hills to Lahore, and then from Lahore to Patiala after the fall of the Sikh kingdom.[19] A painter needed both appreciation and

[17]See my *Pahari Painting* for an analysis of the style of Manak at this point of time.
[18]This group of documents, mostly in Persian, has been published by me under the title *Painters at the Sikh Court,* from the University of Heidelberg (Wiesbaden, 1975).
[19]See, for further details, my *Pahari Painting* on this family and its style.

financial support, and it was natural for him to attach himself to the entourage of a prince or great noble.

There is thus a marked degree of mobility and adaptation, which the painter displays. Quite clearly some of the painters stayed on in their own homes, and suffered a decline in prosperity. Others preferred to move to the Jammu area where the Dogras had founded a kingdom of their own under Maharaja Gulab Singh and his successors.[20] Many of them moved to the plains, as we have seen, in order to be able to survive. Some of them apparently took to doing mural work for the *mahant*s (abbots) of the various religious establishments that lay at the foot-hills, places like Pindori, Damthal and Ram Tatwali.[21] Artists became absorbed in the struggle for a livelihood in which art itself just managed to survive.

The frequent changes in his situation vis-a-vis new patrons made several demands upon the painter. Among them was continuous adjustment to a new range of subjects: the Sikhs for instance were avidly interested in portraits, for these reflected their newly-won glory; their splendour was for them a substitute for dynastic pride. The painter was even earlier wholly capable of good portraiture, which he had done at different periods, and in different styles in the Pahari area: some remarkable work in portraiture belongs to the seventeenth and the eighteenth centuries.[22] But portraiture received new emphasis in the plains, for the Sikh patrons were far less interested in conventional mythological–lyrical themes. Thus the painter, drawing on reserves of skill, concentrated on the task of producing these splendid portraits, some of which are brilliant by any standard. The decline in other works of standard subject-matter was almost made up for by the quality of portraiture at this time. During this period it is not only the formal portraits of the Maharaja and his entourage that impress; the informal, half-finished studies show nearly as much observation and penetration.[23]

Whatever notice was taken by the British of Indian art is marked by a sense of superiority. Ridicule of 'the local Raphaels' is common.[24] They did not understand the conventions of this art, and, fortified with political

[20]Names of Kangra artists like Ruldu and Kanchanu are mentioned at Jammu. There is a drawing with small portrait heads of these artists in the Chandigarh Museum, mentioning them as being *mulazim*s of the Maharaja of Jammu.

[21]Attention to this aspect of patronage of painting by the *mahant*s has been drawn by Karuna Goswamy in her doctoral dissertation, 'Vaishnavism in the Panjab Hill and Pahari Painting' (1968) and, more recently, by Usha Bhatia in her dissertation, 'Painting in the Hindu monastic establishments in the Punjab plains' (1977).

[22]Some very distinguished portraiture was done at Basohli, Mankot and Mandi in the seventeenth century and at Guler, Jasrota and Kangra in the eighteenth.

[23]In fact, some of the informal studies have even greater power than the formal portraiture. Many of them came from family collections of Pahari painters and are now scattered.

[24]G.T. Vigne and von Ujfalvy, who both travelled in the hills in the nineteenth century, make caustic comments of this kind. These are then echoed also in the casual notice that is taken of the work of the Indian painters by the British in the Punjab Hills.

power, made fun whenever they could of these clumsy products of the natives. Worse still, the small *jagir*s which painters had enjoyed as part of traditional payments made to them, were confiscated. The land was returned to them for cultivation but on payment of the usual land revenue. But, for a short while, the British also decided to make some use of the painter's skill. This was for the production of whole series of drawings or coloured sketches showing 'the curiosities of the east', the trades and professions, and the innumerable types of ascetics and holy men, that generally go under the name of 'Company Painting' in India.[25] But this interest was really the product of curiosity and these sets which were often mass produced for the British civil servants could, strangely from the Indian point of view, be bought and sold in the open market. Much of the work was uninspired and only a small part of it was aesthetically valid. Truly speaking, the Pahari painter whether working in the hills or in the plains of the Punjab did not feel comfortable with this distortion of his talent.

In the British period, however, there was yet another, and more serious, complication. The patron's frame of reference, and thus his taste, was very different. Even the rulers of the Indian States, like their English masters, began to favour the more realistic kind of work that had come in with the British.[26] The grandiose oil painting, reflecting all the glory and some of the 'savage splendour' of an eastern potentate's court, was dramatic in its appeal and far more impressive than the tiny miniatures which could neither be displayed to advantage in a darbar hall nor portray photographically the ruler with this debased taste.

The surviving sketchbooks belonging to the artists working in the Sikh states in the second half of the nineteenth century are revealing. The patrons put them to the task of copying elegant designs from European fashion magazines, producing in their own technique the effect of oils and etching, and performing tricks of a similar kind. Some of the painters did make an effort.[27] The move in the direction of realism produced, occasionally, very telling work. But the experiment as a whole did not succeed. The emulation of European painting was beyond their capacity, at least in the short time given them to adjust. The technique was vastly different, the scale of work much larger, the entire outlook radically changed. It is not unlikely that the artist might have risen to the task if other forces had not been simultaneously militating against him. It was the combination of circumstances which led

[25]For a distinguished study of this subject, see M. Archer, *Company Drawings in the India Office Library*, London, HMSO, 1972.

[26]This is one of the saddest chapters in the history of Indian taste. Nearly all Rajas in the nineteenth century preferred these theatrical works to the works done in the earlier tradition by old artists.

[27]A collection of drawings belonging to an old family of artists settled at Patiala, or the remainder of the collections with the Rajol family of artists gives clear indication of the rulers expecting them to do this kind of work.

to his failure. What is of interest to us is that he made a valiant attempt. He took cognizance of a changed situation, produced able and considered responses, and tried to move with the times by enlarging the range of his subject matter and adapting to a new technique and style.

A point which emerges with some clarity from what has been said is the firmness of relationship between patronage and painting in the Pahari area. As soon as patronage shifted, the artist moved; when the level of discernment in the patron declines, art declines. And whatever else this may prove, this firm link between patronage and painting worked both to the painter's advantage and disadvantage. It became a source of strength for him at one time, and, at another, the cause of his decline.

This then appears to be the basic flaw in the system. The common man was very rarely a patron of the art produced in the Pahari area. The ordinary man and woman possessed a feeling for nature, a keen sense of beauty in general, but these manifested themselves in other ways, not in their appreciation of 'high' art, or pride in its ownership. Art was confined to a small circle, restricted within a narrow area.

One of the options that the artist could have taken in the period when princely patronage was on the decline and the artist was being asked to make far too many compromises in his work, was to go to the people and produce work for the people. But this is an option which the painter never took, and possibly it did not even occur to him to do so. Ritualistic paintings had always been done by inferior artists; symbolic drawings on the occasion of marriages in ordinary households or sketches for goldsmiths and embroiderers had always existed. But sophisticated art had never really been brought to the people by court painters.

By the twentieth century, the painters from traditional families had very little work. In order to survive, to begin with many of them tried to make use of their inherited skill and worked as draughtsmen or cartographers in government departments. But this also could not go on for very long. Their skill rapidly declined, and such jobs were in any case comparatively scarce. Within a generation we hear of members of the old painters' families joining the police, the railways and even taking clerical jobs. When I located the artist Chandu Lal of Rajol,[28] he was working as a petty silversmith in his tiny hut using worn instruments and barely making enough to keep his considerable family alive. One of his sons was a mechanic in a road transport company.

A consideration of the developments in the nineteenth century points forcefully in another direction: for good or ill, in the matter of formation of taste, the general populace in the Pahari area took its cue from above.

---

[28]Chandu Lal belongs to the distinguished family of Pahari painters that was headed by Pt. Seu, and is thus a direct descendant of the great Nainsukh. The copies that he makes today based on works by his ancestors contain only faint echoes of the brilliant work of the eighteenth century.

The taste displayed by a ruler and his family communicated itself to the nobility; then to the lower officials, and through them to the people in general. Occasionally this also resulted in bad taste percolating to the lowest level, but fortunately most of the time the taste that was handed down to the people from above was refined and mellow. In any case, the sense of direction in matters of art and taste was not developed by the people themselves but was 'received' by them. The situation has changed very considerably today in other respects but somehow the general public still depends for its values in the matter of taste upon those that are 'above'. The old Sanskrit adage, '*yathā Rajā, tatha prajā*' (like king, like subject'), is wholly applicable also in the matter of art. There are no Rajas any more, but the *prajā* still looks for guidance in taste, and much is left to be desired. Not only is there a general lack of sensitivity, a want of purposeful direction, but there is also bad taste, and patronage without any discrimination.

The vulgarity that one sees becoming increasingly a part of public life in the Pahari area today, especially in the small urban centres, the woeful dulling of sensibilities, is at least in part a general reflection of the fact that good taste is no longer established for people. Such a need for guidance places far greater responsibility upon informed art lovers than is elsewhere the case. In fact, the problem is especially acute in the Himalayan regions where the old kind of art has not yet been replaced either by a new aesthetics which has come in from outside, or by a new response to the changed condition from within the soil itself.

In the Pahari regions as we know them today, thus, there is no visible native impulse in the arts. In the vacuum created by the decay and disappearance of the old art, and in the confusion created by indiscriminate patronage, the whole area seems to be succumbing to philistinism, encouraged by a new-found love of money, the forces of unthinking urbanization, and the industry called tourism which seems bent on sacrificing everything at its own altar.

SELECT BIBLIOGRAPHY

Archer, W.G. 1973. *Indian Paintings from the Punjab Hills*. London: Sotheby, Parke, Bernet.
Francke, A.H. 1914. 1926. *Antiquities of Indian Tibet*, 2 vols. Calcutta.
Goswamy, B.N. 1961. 'The Social Background of Kangra Valley Painting' (Dissertation, Punjab University).
———. 1975. *Painters at the Sikh Court*. Wiesbaden: Franz Steiner Verlag.
Hackin, J. 1931. *La Sculpture Indienne et Tibetaine au Musée Guimet*. Paris.
Khandalavala, Karl. 1958. *Pahari Miniature Painting*. Bombay: New Book Company.
Madanjeet Singh. 1968. *Himalayan Art*. New York: Graphic Society.
Moorcroft, W. & G. Trebeck, 1841. *Travels in the Himalayan Provinces of Hindustan and the Punjab*. Ind. Reprint. New Delhi: Sagar, 1971.
Monod, O. 1954-5. *Peintures Tibetaines*, Paris.

PAL, PRATAPADITYA. 1969. *The Art of Tibet*. New York: Asia Society.
——. 1975. *Nepal: Where the Gods are Young*. New York: Asia Society.
——. 1975. *Bronzes of Kashmir*. New Delhi: Asia Society.
RANDHAWA, M.S. 1954. *Kangra Valley Painting*. New Delhi: Publications Division, Government of India.
SNELLGROVE, D. L, 1957. *Buddhist Himalaya*, Oxford University Press.
SNELLGROVE, D. L. & T. SKORUPSKI, 1977. *The Cultural Heritage of Ladakh*. New Delhi: Vikas.
TUCCI, G. 1949. *Tibetan Painted Scrolls*. 3 vols. Rome.

CHAPTER 10

THE ROMANCE OF SURVEYING IN THE HIMALAYA

K. L. KHOSLA

While the conquerors of the highest Himalayan peaks have achieved deserved renown, few people know anything about the surveyors who mapped the Himalaya, often in conditions of incredible difficulty. They went about their work without any fanfare, giving geographical locations, determining heights and graphically depicting the entire range in easily comprehensible maps. The work of men like George Everest, whose name is commemorated in the highest peak in the world, opened the door to the explorers and climbers who followed.

The earliest surveyors, men like Nain Singh, Kishen Singh and Kinthup, ventured into the unknown with little more than beads to tell their paces. Methods of survey progressed to chaining and the use of the plane table, and, in the latest phase precise theodolites, aircraft and earth satellites mounted with aerial cameras, photogrammetry and computers. Despite the development of such techniques, surveyors still have to climb and explore ranges above 6000 m. These achievements of the surveyors will endure as long as the Himalaya.

EXPLORATORY SURVEYS IN THE HIMALAYA

The development of surveying and cartography is graphically illustrated by two maps. One is a map attributed to Ptolemy (about second century A.D.) where the high Himalaya and some of the major rivers are shown. India as we know it today is hardly recognizable. In a seventeenth-century map attributed to Gerard Mercator (1612), the peninsula is clearly recognizable, but very little information about the hinterland was known to map makers—for the rivers are shown running north to south from a vast lake. The modern physical maps of India are the outcome of two centuries of tireless work.

In 1765, Lord Clive, then Governor of Bengal, deputed Major James Rennell to prepare a map of Bengal. He did it so well that he was appointed the Surveyor General of Bengal in 1767. Surveying for the preparation of maps and charts thereafter became a normal feature of the administration of India. These surveys were mainly confined to the plains, though the Surveyors could not resist making observations of the snowclad mountains in the far distance. They were astounded by the results of their calculations, for they were getting heights far beyond their expectations.

They were also curious to ascertain the sources of great rivers such as the Ganga, Brahmaputra and Indus, which travellers' tales had hitherto described as lakes in the high mountains. Colonel Colsbrooke was anxious to send expeditions to the sources, but all they could do with the funds then available was to make short excursions up the Ganga and Yamuna early in the nineteenth century. After their victorious war with the Gurkhas in 1814, the East India Company annexed present-day Kumaun, Garhwal and Himachal Pradesh. British administration went up into the mountains and the Surveyors followed closely.

Till then, the Andes of South America had been believed to be the highest mountains in the world, but Lambton's Great Trigonometric Series (G.T.S.) described in the next section, had made sufficient progress to prove that the Himalayan peaks were higher. This technique was essentially a series of chains of large imaginary geometric space figures formed by well-chosen peaks and other observation stations all over the country. Measurement of the angles and selected sides of these figures helped to fix the position of the peaks and other stations. The Great Trigonometric Series thus became the framework for the survey of the subcontinent, including the Himalaya.

Webb had already measured Dhaulagiri in 1808–9 in the course of his survey up the Ganga to Gaumukh, and found its height to be over 8000 m. At first this was not believed by western geographers but eventually they had to accept the findings. G.T.S. observations revealed many more peaks of about 8000 m, like Nanda Devi, Annapurna, Makalu, Kanchenjunga, Namcha Barwa, etc. Mount Everest was initially observed merely as a peak. Its computed height led to enquiries being made about its local name. None being found, it was named in 1865 after Everest—the Surveyor General of India from 1830 to 1843—who laid the basis of the Meridional Series and whose work still provides a firm base for present-day surveys. Webb's mission up the Ganga in 1808 raised the question as to whether the Ganga rose from across the great Himalayan Range from the lake Mansarovar. Later Hodgson, who was appointed to survey the former North-West Province (west of Nepal), surveyed up to the source of the Ganga and established that it did not cut through the great Himalayan barrier. In 1811–12 Moorcroft, a veterinary surgeon, and Hearsey made a trip to Mansarovar Lake and discovered that the Satluj actually rose from Rakas-Tal Lake.

In 1814, it was decided to undertake a regular survey of the Himalaya, and the surveys of Crawford and others were extended by surveyors such as Hodgson, Mackenzie, Herbert, Webb and Alexander Gerard, who, with his brother, travelled up the Satluj into Kinnaur and Spiti. In 1821, Moorcroft, Gerard and Trebeck passed through Kangra, Kulu, Lahul and over the Bara Lacha La to Leh. Finally, the three of them pushed across the Karakorams to Bukhara. Tragically, all three lost their lives on the way back.

In this period, survey parties carried instruments like the great theodolite weighing 458.6 kg. (the modern theodolite weighs 18.3 kg) and other camping equipment, up steep and snow-covered mountains without the aid of crampons, ice axes, ropes and other routine mountaineering equipment known today. Kenneth Mason noted in 1860 that a survey *khalasi* (peon) carried a signal pole to the top of Shilla Peak in the Zaskar Range, east of Spiti, at an altitude of 7092 m. 'He did not know its height and we do not know his name.' Such amazing feats were performed by surveyors as a matter of course. Montgomerie, Shelverton and Godwin-Austen were amongst the surveyors who worked in this area and mapped many of its immense glaciers and valleys. It was Montgomerie who made observations on K2, the second-highest peak in the world. Godwin-Austen surveyed the Baltoro Glacier and other approaches to K2, and the glacier at the base was fittingly named after him.

Great changes followed the assumption of control by the British Government in 1857. The Government wanted to know the extent of its territories and the neighbouring states. Thus, from 1865 to the close of the nineteenth century, numerous expeditions were organized in the Himalaya to determine the course of the Indus and the Brahmaputra in their upper reaches. At that time Europeans were not allowed in Nepal and Tibet and other adjacent areas, where Indians otherwise could travel more freely. General Walker, the then Surveyor General of India, therefore decided to train Indians to undertake exploratory surveys in such areas. Captain Montgomerie was in charge of this training and amongst the first to be trained were the now well-known Pandit brothers—Nain Singh and Kishen Singh. They went to Lhasa through Nepal Tibet, and traversed the country between Mongolia in the north-west and Assam in the east, travelling through Ladakh and Towang. They were followed by a host of others like Hari Ram, Kalian Singh, Lal Singh, Mani Singh, Ala Mohammad, Kinthup, Lama Ugen Gyatso, Rinzim Nimgyl, Nem Singh and Atma Ram and Lala; a map showing the extent of some of their travels is illustrated here.

Today, when many of us thoughtlessly pass judgement on maps based on these expeditions, we forget the handicaps the earlier surveyors worked under and the great privations and hardships they suffered. Because of their efforts great gaps in our knowledge of the geography of the region have been filled and for many years little more will be added to this knowledge.

It may be interesting to recall that Kishen Singh was equipped with a 9 inch sextant which was used with the aid of reflections from mercury, a prismatic compass for bearings, and thermometers to measure the boiling

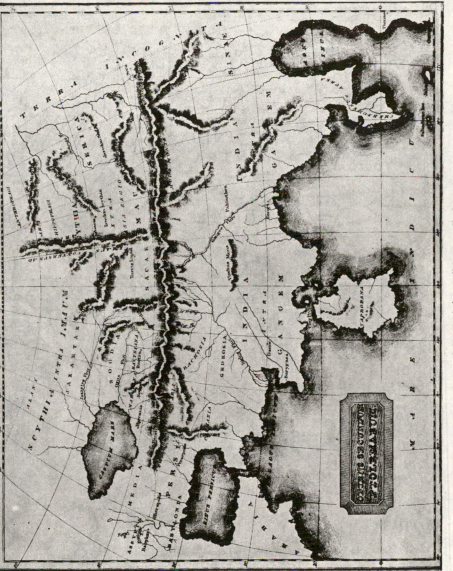

Map of South Asia attributed to Ptolemy.

Map of India attributed to Mercator.

point of water to determine heights. He measured distances by paces and trudged along the Roof of the World without even the luxury of a horse or a pony. When fear of bandits forced Kishen Singh to mount a horse, he first worked out the measure between his own steps and the horse's paces so that he could continue to measure the distance and measurements of the traverse.

For passers-by he was a rider dressed like a lama, in deep meditation, counting his prayer beads. But the surveyor was actually counting paces and recording them on his hundred-bead rosary which was different from the usual Tibetan rosary of one hundred and eight beads. The Buddhist prayer wheel, which he kept turning, was the repository for the beads he set aside to keep track of every thousand paces. In the prayer wheel were also kept other notes on bearings, distances and descriptions. In this manner were recorded the journeys of Kishen Singh in western Tibet and Mongolia as also those of Nain Singh between 1865–75 through Nepal and Lhasa, and from Leh across Tibet to Assam through Lhasa and Towang. Montgomerie, Harman and others who trained the Indian explorers found that their surveys were absolutely accurate.

Kinthup perhaps was the only illiterate amongst them. He actually accompanied a lama who changed his loyalties in the course of their journey, and sold him into slavery. Finally managing to escape, this Sikkimese travelled down the Tsang-po into north Assam's Dihang. When he was only three to four days' march from the plains, he was turned back by hostile tribes. He had unfortunately not thrown any of the marked logs into the Tsang-po, as instructed, and for which Harman's observers waited in vain at Dibrugarh. As a result, the narrative of his travels, despite its being so vivid, was not accepted. Harman himself had died in these four years. However, Colonel Tanner, showing greater faith and perception, recorded it. Thirty years later the expeditionary surveys of Morshead and Bailey, and the Abor expedition of Trenchard, confirmed that the narrative from memory of this illiterate explorer accurately depicted the course of the Tsang-po into the Dihang. Kinthup finally received due recognition and the awards he so richly deserved.

The trans-frontier explorations of Indian surveyors showed that very little was known of our north-eastern frontier. The beginning of the twentieth century saw the organization of a number of expeditions into the unadministered frontier tracts of Assam. Surveyors accompanying these political or military missions were always hard pressed for time in a most difficult terrain, and thus surveying could be done accurately only at a few places and along a few routes.

The expeditions up the Dibang and Dihang valleys clearly established that the Tsang-po enters India as the Dihang (also called Siang) before becoming the Brahmaputra, and that the Dibang does not cross the watershed. The discovery of the Namcha Barwa peak (7829 m) took geographers by surprise as nothing higher than 6154 m was anticipated east of Kanchenjunga, Chomolhari and Kulakangri, thus bearing out an interesting suggestion of Colonel S.G. Burrard's in 1903: 'The Satluj in issuing from Tibet pierces the border range of mountains within 4½ miles of Leopargial, the *highest peak of this region*; the Indus when turning the great Himalayan region passes within 14 miles of Nanga Parbat, the *highest point* of Punjab Himalayas; Hunza River cuts through the Kailas Range within 9 miles of Rakaposhi, the *supreme point* of the range. It will form an interesting problem for investigation whether the Brahmaputra of Tibet has cut its passage across the Assam Himalayas near a point of maximum elevation.' Thus, the Tsang-po from Tibet was finally determined to have curled round the base of a point of supreme altitude, namely Namcha Barwa, to cut through the Assam Himalaya!

The expeditions led by British officers were mainly composed of Indian surveyors. Lieutenant Lewis' expedition into the Miri areas of the Subansiri Valley included surveyors like Yamuna Prasad and Mohammad Nabi besides the *khalasis*. The expedition was greatly impressed by the Apatanis who had terraced rice fields in an area otherwise subject to *jhum* cultivation. They determined that the Kamla and Khru rivers did not cut through the high range visible as the watershed. In 1913, Captain Morshead established that the Subansiri did cut through the high range and had its origin further north in a stream, Charchu, which Pandit Nain Singh had discovered more than thirty years earlier. Much scientific work was carried out by survey parties under Major Osmoston to map the great Himalayan and Zaskar Ranges and the high glaciers and mountains. Eric Shipton in 1936 helped him to reach the Nanda Devi Basin to complete the survey fully by Terrestrial Photogrammetry.[1]

THE SURVEY OF INDIA AND THE GREAT TRIGONOMETRIC SURVEYS

The story of the Himalayan surveys would be incomplete without a brief account of the agency which carried out these surveys and the story of its achievements is synonymous with the successful exploration of the Himalaya.

Survey of India

A brief account of the techniques of the Survey of India, which is the oldest scientific department of the Government of India—it was first set up by Lord Clive in 1767 as the Survey of Bengal—is relevant to the story of Himalayan surveys. The spread of its activity across the subcontinent from Kanyakumari to the great mountains of the north is fittingly embodied in its motto 'A Setu Himachalam'.

The great trigonometrical survey already mentioned provided a frame-work of precise geodetic control points on which our total mapping is based.

[1]Technique of mapping from photographs taken from the ground.

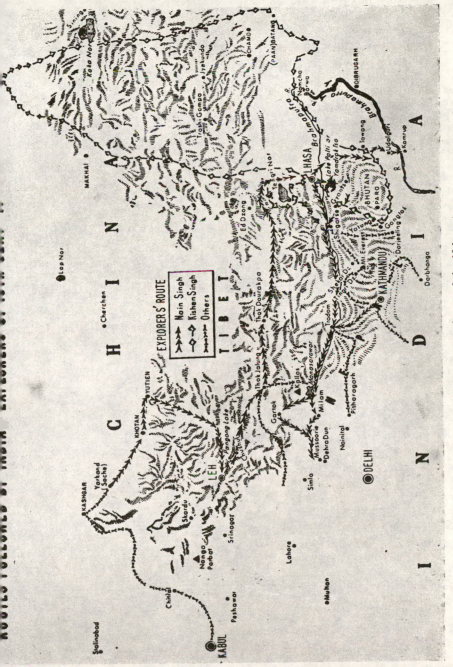

Routes followed by Indian explorers of the 19th century.

These points are the vertices of a set of triangles, quadrilaterals and polygons along meridional (north–south) or longitudinal (east–west) axes. The calculations are relatively less complex operations (in geometry) but the laying of the chains and determination of the control points is a work of immense magnitude, first started at the time when the science of geodesy was in its infancy and the instruments often very cumbersome. The year 1830 marked a further development when a north–south chain of triangles, called the great meridional arc, was measured to evaluate the size and shape of the earth. This work was a major achievement during Colonel Everest's tenure of office as Surveyor General.

The great Himalayan peaks are visible from the principal trigonometrical stations of North Eastern Longitudinal series, and were fixed by measurement with the great 36 inch theodolite. The primary difficulty of the computers who calculated the heights of these peaks was the identification of numerous points, the positions of each of which had been observed from different *view points*. The series was carefully projected on a scale of four miles to an inch, the rays emanating from the stations of observation exactly drawn. Their intersection defined the points sought.

In 1868, the Superintendent of the Great Trigonometrical Survey decided to extend the triangulation through the Province of Assam. It followed the course of the Brahmaputra for a distance of over 480 km, right through the Assam valley to a distance of about 24 km beyond Sadiya, east of the confluence of the three main heads of the Brahmaputra, which here takes an abrupt bend to the south-west. This triangulation is known as the Assam Valley series; it was begun in 1868–9 and completed in 1877–8.

This triangulation has determined the position of a considerable number of peaks in the great snowy ranges of the eastern Himalaya and the mountains to the east of India's extreme north-east frontier. It has also fixed several peaks in the Garo, Khasi, Jaintia, Naga and Singpho Abor Hills which form the southern watershed of the Brahmaputra.

Some of the great surveyors who superintended the most difficult and important work of the Great Trigonometrical Survey were Lambton, Everest, Waugh and Walker.

Mount Everest

Observations to Mount Everest require special mention. The surveyors of the Survey of India Department have observed vertical angles to many high Himalayan peaks. The highest peak of the world, initially designated as Peak H and then Peak XV and finally named Mount Everest, was first observed by a Survey of India officer named Nicholson in 1849–50. The observations were carried out from stations in the plains of Bihar with an average height of 70 m above mean sea level and using towers about 6 to 10 m high built for observations. The observation stations were about 176 km away from the mountain. At the time of observations, the observer had not the slightest idea that it was the highest peak of the world. It was not till 1852 that the computations were sufficiently advanced to indicate that Peak XV possessed a height greater than that of any other known mountain. The peak was neither observed by Colonel Everest himself nor was it observed at the time when he was the Surveyor General of India. Finding a name for this peak then became a matter of paramount importance. From 1852 to 1856 much thought was given to the question, but none of the suggested local names was found acceptable. Consequently, Colonel Waugh, the Surveyor General of India, with the concurrence of Colonel Henry Thullier, Deputy Surveyor General and Radhanath Sikdhar, the Chief Computer, and in consultation with the Royal Geographical Society, finally decided to name it after Sir George Everest (who had actually retired in 1843) to commemorate his contribution to the Geodetic Survey of India. The policy of the Survey of India has always been to adopt local names for all geographical features rather than give them any personal names. Mount Everest is the only exception as no local name was known at the time of its discovery.

Mount Everest was re-observed by officers of the Survey of India during 1952–4 from stations close by at distances varying from 50 to 65 km, using Wild T2 theodolites. The heights of these stations from which Mount Everest was re-observed vary from 2643 to 4499 m. The height of Mount Everest was deduced as 8848 m, and this height has been accepted and published in all the latest maps published by the Survey of India.

Heights of High Himalayan Peaks

The exact heights of the celebrated Himalayan peaks, all over 7000 m, may be of interest:

Peak	Height	
	Feet	Metres
1. Mount Everest	29,028	8848
2. K²	28,250	8611
3. Kanchenjunga I	28,208	8598
4. Lhotse	27,923	8511
5. Kanchenjunga II	27,843	8487
6. Makalu I	27,824	8481
7. Dhaulagiri I	26,810	8172
8. Manaslu I	26,760	8156
9. Cho Oyo	26,750	8153
10. Nanga Parbat I	26,660	8126
11. Annapurna I	26,504	8078
12. Gasherbrum I	26,470	8068
13. Broad Peak I	26,400	8047

Peak	Height	
	Feet	*Metres*
14. Gasherbrum II	26,360	8035
15. Gosainthan I	26,291	8013
16. Gasherbrum III	26,090	7952
17. Annapurna II	26,041	7937
18. Broad Peak II	26,017	7930
19. Gasherbrum IV	26,000	7925
20. Gyachung Kang	25,910	7897
21. Disteghil Sar I	25,868	7885
22. Himal Chuli	25,801	7864
23. Kangbachan	25,782	7858
24. Ngojumba Kang	25,720	7839
25. Nuptse	25,719	7839
26. Manaslu II	25,705	7835
27. Masherbrum E	25,660	7821
28. Nanda Devi	25,645	7817
29. Kamet	25,447	7756

MODERN GEODETIC AND GEOPHYSICAL SURVEYS

Modern surveying activities go well beyond the explorations and the Great Trignometric surveys described earlier. The rapid development of modern surveying instruments and techniques of observation and analyses of data have played an important role in the evolutionary process and have made their impact on the progress of survey operations in India. Modern surveying activities in the Himalaya are of two distinct kinds.

The geodetic and geophysical surveys in the Himalaya cover the following activities with the highest precision: (a) Horizontal control surveys; (b) Vertical control surveys; (c) Astronomical surveys; (d) Gravity surveys; (e) Geomagnetic surveys.

Horizontal Control Surveys

Horizontal control surveys aim at fixing the latitude and longitude of some pre-selected points (called stations) which form the basis of further surveys. Observation of series of triangulation chains, described earlier, was the main procedure used for this purpose in the Himalayan region.

The bulk of the triangulation series had been completed during the nineteenth century and these provided a network of strong chains of triangles covering the entire length and breadth of the Indian subcontinent. The only extension of this work in the Himalaya was the north-west Himalaya series; Kashmir Principal; Kumaun and Garhwal and Assam longitudinal series, to mention a few. Subsequently, Geodetic Triangulation has also been executed for the Chittagong Series and Assam Longitudinal series extending to the Burma border.

Although geodetic triangulation has not been extended in the interior of the Himalaya, excepting Ladakh, geodetic control in the foot-hills has been utilized for execution of large-scale topographical triangulation, notably in Nepal, Bhutan and Arunachal Pradesh, for mapping requirements during the 1960s and the early 1970s.

The theodolites used for measuring angles in this triangulation (up to 1930) were the 12, 18, 24, and 36 inch models[2] which were very bulky and cumbersome. However, by evolving an elaborate system of observational drills, the accuracy of work was not allowed to suffer. The sophisticated glass arc Geodetic Theodolite Wild T3 was introduced for the first time in 1930 and used for work on the Chittagong Series. Thereafter this type of theodolite has been used exclusively for all geodetic triangulation. The most sophisticated Universal theodolite (WILD T4) was acquired by the Department in 1958 and this has also been used where triangulation had to be combined with astronomical determination of longitude and latitude. After the last war there was a revolutionary change in geodetic instrumentation as a result of the development of electronics. Distance-measuring devices using microwaves and laser/infra red propagation of energy, are now available. New procedures for fixing Geodetic Control called 'Trilateration[3] and High Precision Traverse' were developed and greatly facilitated rapid work. The instruments used in such procedures are tellurometers, distomats and geodimeters. Geodimeters are now exclusively used for base measurements.[4]

The latest technique for provision of control is now available with us in the form of Doppler receivers. In a matter of a few days the position of any point in the Himalaya terrain can be obtained by receiving transmissions from earth satellites.

Vertical Control Surveys

Vertical control surveys aim at fixing the elevations of some pre-selected points (called bench marks) above the sea level. This is achieved by observations with a spirit level which gives a horizontal line of sight and can determine the relative elevation difference between any two points at which graduated staves are held vertically for observation. This procedure called levelling was used to establish a network of bench marks.

The first level net carried out between 1858–1909 was adjusted in 1909.

---

[2] The sizes indicate the size of the horizontal circles on the theodolite on which engravings for measurement of horizontal angles are marked.

[3] The three sides of the triangle are measured directly.

There are the selected sides for which most precise linear measurements are carried out for triangulation computations and adjustments.

Although a fair amount of coverage was provided, all the levelling was con-
fined to areas outside the Himalayan belt except the line Dehra Dun to
Mussoorie observed in 1861–2. Later, a few levelling lines were extended
into the Himalaya, such as Bareilly–Naini Tal, Siliguri–Darjiling and
Rawalpindi–Srinagar. Srinagar was again connected with Jammu, and Dehra
Dun was connected to Mussoorie via Chakrata.

With the development of communications into the inner reaches of the
Himalaya, and development of the water resources of the region, intensive
levelling work has been carried out during the five-year plans. Repeat
observations cf levelling have also provided clues about relative earth-surface
movements in fault and thrust areas.

Astronomical Surveys

Astronomical surveys aim to determine the latitude and longitude of some
pre-selected points by observations to celestial bodies like stars.

During the earlier surveys, a few latitude stations were established in
Kashmir south of the Indus and a few north of Dehra Dun and around
Darjiling. During 1952–77, stations were established by astronomical
observations in the Himalaya and its foot-hills in connection with the new
determination of the height of Mount Everest and adjoining peaks.

With the acquisition of the Wild T4 Universal theodolite, the T3 astrolabe
attachment and portable dry battery wireless sets during the late fifties,
field operations for astronomical work have been greatly facilitated. It is
now possible to cover hitherto inaccessible high-altitude areas with ease,
thus speeding up astronomical work in Himalayan regions.

Gravimetry

The gravity force of the earth is essentially the sum total of the attracting
force due to its mass and the centrifugal force due to its rotational motion,
which it exerts on any particle coming within its sphere of influence. All
the measurements of the surveys, be it with a level or a theodolite, are
carried out in relation to the direction of the vertical which is none other
than the direction of gravity. The study of the earth's gravity field has been
a fascinating task for surveyors. Gravity observations are conducted for
the determination of the shape of the earth and other geodetic studies.
They also provide valuable data for prospecting for minerals including oil.

Prior to 1947, the determination of gravity was done by means of the
usual pendulum observations. This method was slow and cumbersome and
was snagged by uncertainties. Accuracy was of a low order. Out of a total
of over 400 pendulum stations only about 22 were located in the Himalaya,
mostly in Kashmir and a few around Darjiling. During 1947, portable
instruments called gravimeters, compensated for pressure and temperature

changes, were developed and one such instrument, namely the Frost gravi-meter, was acquired by the Department. Later there was further improve-ment in the design of these instruments and much more precise and light models were acquired. It is now possible to establish a 15 km-mesh of gravity stations all over the country. Gravity studies combined with the astro-nomical studies made by surveyors, particularly in Himalayan areas, have helped to establish the existence of isostatic compensation. India is known as the birth place of isostacy, a theory according to which the earth's crustal elements (whether hills or valleys) are so constituted that they exert the same pressure at a depth not too far below the sea level.

Geomagnetism

It is common knowledge that the earth behaves as a magnet, and magnetic compasses have helped explorers to obtain directions. But the directions given by the compass vary from true north, and magnetic surveys are neces-sary to study such variations for help in accurate navigation and other scientific investigations.

The immediate aim of modern magnetic surveys is to determine the earth's magnetic force in magnitude and direction in every portion of the area involved and to measure the changes in this field. These surveys also reveal areas of magnetic anomalies or disturbances and enable further investiga-tion of such areas to be made. For data for compilation of charts to be obtained, a number of permanent 'Repeat Stations' are established all over the country at intervals of 50–80 km where periodic observations are conducted.

Prior to 1962, there were no magnetic repeat stations located in the Himalaya except one at Darjiling, which left a big void in the north. How-ever, after 1962, 29 magnetic repeat stations have been established in the Himalayan region, where periodic observations are conducted as a regular feature. Permanent magnetic observatories are functioning at Sabawala, Shillong and Gulmarg in the Himalaya and their foot-hills.

MODERN METHODS OF SURVEYING AND MAPPING

The geodetic surveys described so far provide the basic framework for the detailed surveying and mapping of hills and valleys, lakes, forests and other natural and man-made features.

A topographic map is a graphical representation of selected natural and man-made features of a part of the earth's surface on a definite scale. The features are depicted in their correct geographic positions and elevations. Symbols and colour differentiations and contours help to show the physical features, mountains, valleys and plains in their true relationship to the surrounding man-made and other natural features; in a way, they are an inventory of the physical features on the surface of the earth.

Earlier topographic surveys necessitated work being done wholly on the ground, using theodolites and plane tables. This was however possible only in areas where man could function. Peaks over 6150 m were extensively climbed and surveyors moved and lived for long periods in the most difficult terrain and inhospitable environment, away from civilization, family and friends. The process was also time consuming.

The advent of *aerial photography* and *photogrammetry* revolutionized surveying. It enabled surveyors to carry out the larger part of survey work from the cosy comfort of their offices. Though the romance with the Himalaya has not ended—surveyors still have to go to the field to carry out the minimal survey work—a very large part of agonising effort has been eliminated.

Of the various types of photographs used in surveying, aerial photographs, taken with the axis of the air-borne camera vertically downwards or nearly vertical, are now mainly relied upon for topographical mapping.

Photogrammetric Procedure

Topography constitutes a three-dimensional physical model. If aerial photographs of this model are taken with a camera carried in an aircraft, with suitable overlaps between successive pictures (usually 60 per cent), it is possible with the help of these photographs and suitable photogrammetric machines to recreate a measurable visual model of topography, on a smaller scale, in our offices.

The overlapping pictures enable the same ground to be recorded as observed from two viewpoints along the flight path. These images would correspond to the two images formed by the two eyes on our retina, which are 65 mm apart, to provide a single three-dimensional image of all that we see. Viewing the two recorded aerial pictures of the same area by our two eyes (one picture by one eye at a time) recreates in a way the two ground images on our retina, which are physiologically merged by 'stereoscopic' vision to form a three-dimensional view of the ground. This makes it possible to bring the terrain to the table of the surveyor in its three-dimensional form—seen as it would appear to a giant—as tall as the height of the aircraft from which the photographs were taken and with eyes as far apart as the two successive positions of the aircraft from which the two consecutive pictures were taken.

Having thus brought a faithful model of the Himalaya to the office we can carry out our surveys in cushioned comfort, though we may miss the thrill of glimpses of snow peaks, which would appear only as a dull grey or white patch on the model recreated from the photographs.

It is analogous to viewing a reduced model of ground from very near, and we can now measure accurately, vertically and horizontally. We can determine the relative heights of objects and their relative plan positions.

One can appreciate the convenience of carrying out Himalayan surveys of lofty peaks like Mount Everest, K2, Kanchenjunga, steep gorges, deep valleys and forbidding glaciers in such a way. Today *photogrammetry* is recognized as a basic procedure in all types of terrain mapping and is being increasingly used in numerous other fields.

Photogrammetry in the Survey of India

Starting as early as 1899, the Survey of India lost no time in picking up this new technique. At first, photographs were taken from the ground and a number of creditable surveys were made using terrestrial photography. Eventually, however, because of unfavourable terrain, this method was superseded by aerial photography.

A beginning with aerial photography was actually made by the Survey of India in 1916 on the Mesopotamian front in the First World War. Encouraged by the results, the Indian Air Survey Committee was formed in March 1927 and air surveys were greatly extended during the Second World War, for revision of existing maps as well as the preparation of new ones.

The next major departure was in 1948 when Brigadier Heaney, Surveyor General, with the assistance of Colonel Wilson and the advice of Dr Hart, initiated a period of training development. Since 1964, the Survey of India has used the latest highly-sophisticated instruments for aerial photography, a development that has greatly facilitated survey and mapping, particularly of mountainous regions.

Modern Trends

The department now has 14 fully-equipped photogrammetric units with a large complement of technicians and engineers, with a wide variety of the latest instrumentation. Further sophistication has resulted from the introduction of computers in 1970.

Aerial photographs are now being increasingly used in the survey for evaluation of our natural resources in the Himalaya—a region which has remained largely untapped. Mapping of the extent and depth of snows, the movement of glaciers, and scientific studies for the conservation of soil and forest resources, and of flora and fauna for the maintenance of ecological balance, are being increasingly undertaken using aerial photographs. Satellite technology has now made it possible for us to study land and water resources and other features even beyond national frontiers. Detailed mapping of inaccessible regions which would have earlier taken perhaps a hundred years, can now be completed in the relatively short span of a decade.

While this advanced technology has changed the whole character of

Himalayan surveying, it should not be forgotten that the foundations were laid as early as 1814 by devoted men who often endured extreme privation to enable us to grasp the majesty of the Himalaya from the safety of our rooms. So long as man's quest for knowledge remains, the present can at best be viewed only as a phase in the continuing story of 'The Romance of Himalayan Surveys'.

BIBLIOGRAPHY

AGGARWALA, R.K. 1971. 'Deputation Report on the Ist International Workshop on Earth Resources and Survey Systems'. VII International Symposium on Remote Sensing. Ann Arbor, Michigan, USA.

BRUCE, C.G. 1934. *Himalayan Wanderer*. London: Alexander Markhouse.

BURRARD, S.G. 1915. *Records of the Survey of India*, vol. VIII.

—— and HAYDEN, H.R. 1933. *A Sketch of the Geography and Geology of the Himalayan Mountains and Tibet*.

BURRARD, S.G. *Survey of India, Professional Paper, No. 9* (1905).

DAS, S.C. 1904. *Journey to Lhasa and Central Tibet*. London: John Murray.

DREW, F. 1975. *Jummu & Kashmir Territories: A Geographical Account*. London: Edward Stanford.

——. 1912. Panoramas Maps & Index. London, Constable and Co. Ltd.

GULATEE, B.L. 1954. *Survey of India. Technical Paper Nos. 4* (1950), 8.

CUNNINGHAM, A. 1871. *Ancient Geography of India*. London: Trubner.

HEDIN, SVEN. 1909–13. *TransHimalayas: Discovering Adventures in Tibet*. London: Macmillan.

HENNESSEY, J.B.N. 1891. *Report on the Explorations in Greater Tibet and Mongolia, Made by A-k in 1879-82*.

HOOKER, J.D. 1854. *Hooker's Himalayan Journal*. London: John Murray.

KARAN, P.P. 1963. *Himalayan Kingdom: Bhutan, Sikkim & Nepal*. Toronto: D. Van Nostrand.

LONDON, P. 1928. *Nepal* vols. I & II.

MARKHAM, C.R. 1978. *A Memoir on the Indian Surveys*.

MASON, KENNETH, 1955. *Abode of Snow*. London: Rupert Hart-Davis.

MOORCRAFT and TREBECK 1841. *Travels in the Himalayan Province*. London: John Murray.

CHAPTER 11

MOUNTAINEERING IN THE HIMALAYA

H. C. SARIN and GYAN SINGH

EARLY TRAVELLERS

Since the dawn of civilization Indians have been drawn to the Himalaya. Many holy men crossed high and hazardous ranges and established places of pilgrimage far from cities and villages at places like Kailash, Badrinath, Kedarnath, Gangotri, Jamnotri, Amarnath, and Hemkund. Though the mountains were remote, pilgrims journeyed in numbers, with no special aids, seeking religious fulfilment. The harder the journey and more austere the food, the nearer it brought them to Divinity and *Moksha* or salvation. To treat these journeys lightly was considered sacrilege. Perhaps the most notable pilgrim of all was Sankaracharya who a thousand years ago came from south Kerala to revitalize Hinduism and establish a shrine at Badrinath near the source of at Alaknanda.

It was primarily economic compulsions which, much later, drove people living in the foot-hills of the Himalayan range to accept travelling in the mountains as a way of life. As traders, hunters, shepherds and porters they had to cross mountain passes as high as 5538 or 5846 m, and the need for designing and fabricating indigenous equipment for negotiating snow and ice slopes became apparent to them. Some of these items are still in use, e.g. *lham,* all-wool thickly woven boots which reach up to the knees and are warmer and more comfortable on ice and snow than ordinary leather boots; a walking stick with a ferrule which helped maintain balance on tricky slopes; yak hair rope for climbing, used mostly as fixed rope; jackets made of lamb-skin, and long bamboo poles for negotiating crevassed glaciers; strands of their long hair were drawn across their eyes to protect them from glare and snow blindness. Several thousand traders would wait every year for the snow to melt on border passes, and then proceed with goats, sheep and yak laden with foodgrains and other products from India and return from Tibet with wool, salt, horses, hides and skins before the passes closed.

'The world's altitude record as far as we know', wrote Kenneth Mason, 'was held for about twenty years by one such Khalasi engaged by the Survey of India, on a salary of six rupees a month (about 12 shillings) who carried a signal pole in 1860 to the top of Shilla, in the Zaskar range east of Spiti, 23,950 feet above the sea. He did not know his height and we do not know his name.' We now know that the height is actually 6191 m.

The most famous of the Indian explorers of the time in the Survey of India (1865–85) were the Pandit Brothers—Nain Singh and Kishen Singh. They were sent by the British to the cold desert of Tibet and adjacent countries. Nain Singh's 2400 km long journey from Ladakh to Lhasa via Changthang in 1865–7, and Kishen Singh's four-year journey (1878–82) from Darjiling to Outer Mongolia and back, involving several thousand kilometres of strenuous trekking over the mountains, is a saga of mountain travel. Hari Ram was the first to make a circuit of the Everest group in 1871. He was arrested for suspicious activities, but secured his release by curing the ailing wife of a local notable, though he was not a doctor. Das in the East crossed the Jonsang La (6154 m) into Tibet in the eighties of the nineteenth century. The surveys made by these pundits were based on the counting of steps through prayer beads, and were amazingly accurate. They earned great praise from those that followed them, including the then Surveyor-General, who referred to Kishen Singh as 'accurate, truthful, brave, and highly efficient'.

THE BEGINNING OF MOUNTAINEERING IN THE HIMALAYA

Mountaineering as a sport started in the Himalaya in 1883, with the climber W.W. Graham, who was the first European to come for the sole purpose of climbing for sport. Other Europeans had indeed visited these mountains much earlier with a variety of objectives. In 1624 two Portuguese priests, Father Antonio de Andrade and Brother Manuel Marques, started on a missionary assignment from Agra ard went to Tsaparang on the upper Satluj in Tibet. Travelling up the Alaknanda Valley and moving north via Badrinath, they crossed over the 5661 m Mana Pass. Nineteenth century Himalayan adventurers in pursuit of political objectives, were William Moorcroft, Victor Jacquemont, Joseph Wolff, Baron Carl von Hugel, G.T. Vigne, Colonel Alexander Gardiner, Alexander Burness, and George Hayward. Most of them worked for the East India Company and journeyed from Ranjit Singh's dominions in Panjab to Himalayan regions covering Gilgit, Ladakh, Afghanistan, and even dared to venture into Tibet, crossing high ranges, facing mountain hazards, often suffering imprisonment and even making the supreme sacrifice in the name of the 'Great Game'—a euphemism for the expansion and consolidation of British power.

One of the most valuable scientific explorations was undertaken towards the middle of the nineteenth century by the famous botanist and naturalist, Joseph Dalton Hooker. He spent two years, in 1848 and 1849, in the Sikkim Himalaya and gave detailed accounts of the flora of Sikkim, and of his travels in what he called 'Himalayan Journals'. These accounts were instrumental in focussing the attention of the western world on the richness of the Himalayan flora.

To revert to W.W. Graham; he went first to the Sikkim Himalaya with two Swiss guides in the spring of 1883, and after some reconnaissance in this region, went to Kumaun in July, where he tried to approach Nanda Devi but failed to negotiate the Rishi Gorge. He attempted Dunagiri (7134 m) and later claimed to have climbed Changabang (6929 m). Graham returned to Sikkim where he claimed that he had climbed Kabru (7385 m). His claims were critically examined by experts in later years and were generally not accepted. He did not have any instruments and was, at best, an indifferent map reader. His exploits, however, attracted much attention among European alpinists.

At the turn of the century three areas attracted western mountaineers, namely the Karakoram, Kumaun and Sikkim Himalaya. In the Karakoram, this period saw a large number of mountaineering explorations even though no peak worth the name was climbed. The Duke of Abruzzi and Filippo Da Filippi reconnoitred the approaches to K2 and extensively explored many glaciers in the region. In Kumaun Himalaya, Dr Longstaff explored the approaches to Nanda Devi in 1905. Later, in 1907 he made the first ascent of Trisul (7188 m), the highest summit reached till then. Another mountaineer, C.F. Meade, made three determined attempts on Kamet (7830 m) in 1910, 1912 and 1913, but could not climb higher than a pass which was later named Meade's Col (7203 m).

Some very valuable mountaineering explorations were carried out in the Sikkim Himalaya by Major L.A. Waddell, Claude White, and Freshfield in the last years of the nineteenth century. These eminent mountaineers climbed extensively around Kanchenjunga and its satellites. They also wrote informative books which are still of interest to mountaineers who wish to climb in the region. Two Norwegians, C.W. Rubenson and Monrad Aas, made a gallant attempt to climb Kabru in 1907, but devastatingly cold winds robbed them of success when they were barely a few metres short of their goal.

Dr A.M. Kellas of Glasgow first came to Sikkim in 1907. He seemed to have enjoyed his climbs in this region as he returned in 1909, 1910, 1912 and, after the First World War, in 1920 and 1921. He climbed in Kumaun in 1911 and 1914. His visit in 1910 was particularly rewarding; he scaled Langpo Peak (6923 m), Sentinel Peak (6902 m), Paunhri (7132 m) and Chomiomo (6902 m). Apart from his mountaineering achievements, Dr Kellas was perhaps the first expert to carry out studies on the physiological problems at high altitudes.

In the more leisurely and carefree days of the nineteenth century and the years before the First World War, practically the entire Himalaya lay unexplored and hundred of peaks of 6154 m and above were yet to be attempted. The influx of explorers, surveyors and mountaineers had no real effect on the resources or the ecology of the Himalaya. The local population which always had the right to forest wealth were allowed to clear as much forest as they needed. Government authorities were generous in recognizing such rights. It is not surprising that an educated and enlightened man like Joseph Hooker watched forest fires with the same excitement and delight as a child witnesses a display of fireworks. In one of his Journals of 1848 he wrote, 'At this season the firing of the jungle is a frequent practice, and the effect by night is exceedingly fine; a forest so dry and full of bamboo, extending over such steep hills afforded grand blazing spectacles. The voices of birds and insects being hushed, nothing is audible but the harsh roar of the rivers and occasionally, rising far above it, that of the forest fires.'

BETWEEN THE WORLD WARS

In the two decades after the First World War, continental climbers made headway not only in some notable ascents in the Alps, but also in the development of mountaineering techniques. About this time the Himalaya began to attract more and more western alpinists. Some nations, however, stuck to the mountains they had originally selected as their objectives: for example, while the Germans concentrated mainly on Nanga Parbat and Kanchenjunga, the British seemed to have made Everest a national challenge. The Americans, on the other hand, considered K2 as their mountain. A notable and healthy departure from the pattern was the joint British–American expedition led by T. Graham Brown which climbed Nanda Devi in 1936.

Everest

Everest has always held a spell for mountaineers. Since the ascent in 1953 by Tenzing and Hillary, it has been climbed many times, yet it remains the biggest challenge of all Himalayan giants. Our knowledge of Everest is popularly believed to have started dramatically in 1852, when the computer of the Survey of India, Radhanath Sikdhar, rushed into the office of his Surveyor-General in Calcutta and announced breathlessly: 'I have discovered the highest mountain in the world.'

The height of Everest, 8924 m (later, allowing for atmospheric refraction, 8848 m) was computed from a mass of figures representing triangulation survey observations carried out from the plains of India in 1848, three years earlier. Its humble name, Peak XV, was changed to Mount Everest, after Sir George Everest, a distinguished Surveyor-General who was largely responsible for the trigonometric surveys of the Himalaya. The Tibetans, however, continued to call it Chomolungma (Mother Goddess of the winds). The Nepalese call it Sagar Matha (Forehead of the Seas).

Nepal and Tibet, the two countries through which Everest could be approached, were closed to foreigners. No organized party could, therefore, go near this great mountain for nearly seven decades after its discovery. However, there were a few surreptitious attempts at reconnaissance of the base of Mount Everest through Tibet as well as Nepal, but the First World War ended all climbing for sport.

After the war, relentless pressure on the Dalai Lama, through British diplomatic sources in Lhasa, yielded results and permission was at last granted to foreigners to explore the Everest region. Explorations started in earnest in 1921. The first expedition organized in 1921 under the auspices of the Royal Geographical Society and the Alpine Club was led by Colonel Howard Bury. Members of the expedition surveyed over 38,000 sq. km in three months and produced reasonably clear maps of the region. On this tedious reconnaissance trip, Mallory was the soul of the mountaineering group which reached the North Col (about 7880 m). Scaling Everest became an obsession for him, which tragically ended his life in 1924.

The second Everest expedition in 1922, brilliantly led by General Bruce, made a remarkable advance. One cannot but admire the enthusiasm and tenacity of the members of this venture. Mallory, Norton and Somerville reached a height of 8307 m without oxygen. Later, two other climbers, Finch and Bruce (General's cousin) reached an altitude of 8380 m using rather indifferent oxygen equipment. The expedition had to be called off when, on 7 June 1922, Mallory's party was caught in an avalanche which killed seven Sherpas.

The historic third expedition (1924) was led by Norton and generally followed the same plan and pattern as the 1922 attempt. Norton and Somerville were able to ascend to a height of about 8646 m without oxygen. On 8 June Mallory and Irvine, using oxygen, climbed towards the summit. This intrepid pair was last seen at about 8615 m. They could conceivably have attained their objective but they never returned to tell their story. Nine years later, the climbers of the fourth expedition found one ice axe in a state of perfect preservation at a height below the point where the two mountaineers were last seen by Odell.

After this tragedy, for eight years the Dalai Lama forbade further expeditions. This was a terrible blow for the British because Everest had become almost a national challenge. Since the Dalai Lama was adamant, they made an official request to the Maharaja of Nepal to let them attempt the mountain from the Nepalese face. Failing in Nepal, they went back to the Dalai Lama and through diplomatic persuasion obtained permission for 1933.

The fifteen-man expedition of 1933 led by Hugh Ruttledge included two famous mountaineers, Frank Smythe and Eric Shipton. Since the expedition was dogged by bad weather from the very beginning, it could make no headway beyond about 8615 m. In 1935, Shipton's lightweight (fifth) team of six climbers and fifteen 'Tiger' Sherpas was only a team for 'reconnaissance in force'. Although they reached the North Col in record time, the weather had already deteriorated, preventing them from going further. Before returning to Darjiling, however, they climbed a number of lesser peaks in the area. Shipton also climbed up to Lhola and had a peep into the Khumbu Basin on the southern side of Everest. Like Mallory, he declared the southern approach to Everest to be impracticable. The next year's attempt by practically the same group was led by Ruttledge. Bad weather combined with avalanche-prone conditions thwarted all attempts to climb to any significant height. It was not for the first time that mountaineers were made to realize that it is not only physical strength and climbing skill and experience which count; the weather holds the trump card.

After his success on Nanda Devi, Tilman was chosen to lead the seventh attempt in 1938. By temperament an ascetic, Tilman sacrificed everything he considered superfluous, and his highly streamlined expedition had only seven climbers and forty-five Sherpas. He permitted no reserves in man-power or logistics. However, despite very heavy snow, Shipton and Smythe climbed up to 8400 m. One more abortive attempt, and the expedition had to retreat.

The Himalayan Club

The Himalayan Club, founded in India in 1927–8, became a symbol of the development of mountaineering in the Himalaya. Field Marshal Sir William Birdwood, the British Commander-in-Chief in India, was the Founder President, and the club prospered under the patronage of the Viceroy, Governors, and other high officials, and the fine traditions established by the leading climbers of the time.

The Himalayan Club's aim, based on that of the British Alpine Club, was declared to be 'To encourage and assist Himalayan travel and exploration and to extend knowledge of the Himalaya and adjoining mountain ranges through science, art, literature and sports.' In 1929, the Club started publishing *The Himalayan Journal* which in format and other details was modelled on the *Alpine Journal*. The Club helped expeditions from abroad, and later even hired out mountaineering clothing and equipment to its members at reasonable rates. The Club also compiled a list of experienced Sherpas and instituted the award of the 'Tiger Badge' for outstanding performance by Sherpas. It took measures for the welfare of Sherpas like laying down rates of payment, scales of rations, and equipment, as well as the question of compensation in the event of disability or death. The Himalayan Club acted as a focus for a wide range of activities in the Himalayan range, till the mid-fifties, when large-scale international expeditions began to take place.

Kanchenjunga

In 1929, Paul Bauer brought to the Himalaya a strong team of nine climbers in quest of a really demanding challenge. With the advice and help of the newly-formed Himalayan Club, they decided to attempt Kanchenjunga, the third-highest mountain in the world. This post-monsoon maiden attempt from the Sikkim side eventually proved to be only an elaborate reconnaissance. The team established ten camps, the highest being at 7100 m. They reconnoitred the route up to a height of 7400 m. A very strong blizzard and bad weather for more than three days pinned them down to the ridge; consequently, they had to withdraw and were lucky to come off the mountain without any casualty.

The 1930 Kanchenjunga expedition organized by Professor G.O. Dyhrenfurth was originally meant to be a German venture, but finally developed into an International Expedition with Germans, British, Swiss, and Austrian participants. They approached the mountain by a much longer north-west route in the pre-monsoon period. Regrettably, this expedition could not make much headway. They had established three camps when their most experienced Sherpa, Chetan, was killed in an ice avalanche from a hanging glacier. After this mishap, the team abandoned the route, but reconnoitred another route which also proved unfeasible, and eventually they had to call off the attempt. Nevertheless, before returning they climbed peaks ranging from 6154 to about 7385 m.

In 1931 Bauer returned to his mountain with a strong team of nine, five from his 1929 party, with four new and younger alpinists. His familiarity with the Zemu approach enabled him to cover the ground fast. Seven weeks after leaving Munich, they were at camp IV, but it was already July and the monsoon was in full swing in the eastern Himalaya. On 9 August, tragedy struck the party: Hermann Schaller and Sherpa Pasang were killed when snow gave way under them on the ridge. This mishap cost them fifteen valuable days. In any case, they were able to establish camp XI at 7431 m and climb to a height of 8000 m. Beyond this there was a stretch of powder snow on hard ice, so movement on this part of the route was not only difficult but positively dangerous. They could not risk waiting for the snow to stabilize because one heavy snowfall would have meant disaster. Besides, they had also been at heights in the deterioration zone for over six weeks with hardly any reserves of strength. Bauer wisely decided to withdraw. For most of the group the descent was a harrowing experience. They eventually got back to their advance base on 24 September.

Nanga Parbat

The first expedition to Nanga Parbat in 1932 was led by Willy Merkl who was considered to be the greatest mountaineer of the period. While he had great knowledge of the Alps, Merkl had no experience of Himalayan heights, particularly of the magnitude of logistics, or the problems of weather conditions at heights of 7000 m and above. Merkl had a strong team but the expedition could not go far without any experienced high-altitude porters, and only three out of the eight members could reach 7077 m. The expedition ended as a reasonably successful reconnaissance trip during which a practicable route was found.

Merkl returned to Nanga Parbat in 1934 a much wiser man, but one who still underestimated the physiological problems of high altitudes. In his team, there were nine climbers in all, every one of whom was technically highly competent. This time they had recruited thirty-five experienced Sherpas and Bhotiya high-altitude porters. After they had established the first three camps, one of the members died of pneumonia. Despite some delay on this account the determined climbers went ahead. Camp VII was established at a height of 7105 m. Then Merkl with four climbers and two porters reached the Silver Saddle, 7451 m, and set up camp VIII at 7480 m. That night a raging storm broke out on the mountain, and only two climbers and five Sherpas could reach safety in the higher camps. For five days climbers and Sherpas at camp IV made repeated attempts to reach camp VI, but the blizzards and heavy snow thwarted all rescue efforts. In the nightmarish retreat, Merkl, his two companions and six Sherpas perished. This great tragedy was redeemed by many acts of great heroism and valour. Of "it" Paul Bauer has written: 'Nature, which had marked out Merkl to be a leader, had also endowed him with the physical stamina to hold out to the last in storm and cold. It had been his fate to drain the cup to the bitter dregs. On July 9 he was witness to the death of kind Uli Wieland; four days later, his trusted friend and rope-mate Willo Welzenbach expired by his side; and at the end, he the organizer of a big and well-equipped undertaking, was reduced to a single blanket shared with Gaylay, who might have saved himself by joining Ang Tsering, but who preferred to stay with his master, faithful even to death.'

Thus ended one of the most poignant, though not the greatest mountain disasters of the time: that was to follow. Paul Bauer who was by now a father-figure among German mountaineers, took three younger men to Sikkim in 1936. His aim was not Kanchenjunga; he wanted to train fresh material. They made the difficult first ascents of Siniolchu (6891 m) and Simvu (6607 m).

Wien, one of these three climbers, organized an eight-man team to go back to Nanga Parbat with twelve Sherpas from Darjiling in 1937. The first mishap of the expedition took place at an early stage when camp II was overtaken by an avalanche. There was some damage to the equipment but fortunately no serious casualty. A few days later the team continued and stocked higher camps. They established camp IV in an ice cave at 6240 m. This became their Advance Base and was fully stocked and occupied by climbers and nine porters by 11 July. During the next three days there were spells of heavy snow. They tried unsuccessfully to push forward, but then the pitiless Nanga Parbat struck again. When a ferry of five porters carrying stores led by Luft reached the site, the whole camp was found buried under two-day-old frozen snow. Rescue parties had to come from distant places. But by then Nanga Parbat had taken a toll of sixteen men—seven members and nine porters. Luft was the only survivor from the German team.

Undaunted, Bauer decided to avenge the death of his friends and organized a meticulously-planned expedition to Nanga Parbat in 1938. Their first attempt was frustrated by bad weather just short of the Silver Saddle. The second push could not go even that far. Bauer decided to call off the expedition; the weather was unfavourable all over the Karakoram and would have led to risking lives again.

A very small party under Aufschnaiter reconnoitred the route through the Diamirai gorge in 1939. They reported that a compact party of technically-skilled climbers with a small administrative tail had a good chance of making the summit along this route. However, the Second World War put a stop to all serious climbing in the Himalaya.

Mountaineering in Kumaun

In 1931 Frank Smythe, the famous mountaineer, naturalist, writer and photographer, led a strong team including Eric Shipton and R.L. Holdsworth to climb Kamet (7756 m). The first ascent was made by Smythe, Shipton, Holdsworth and Sherpa Lewa on 21 June 1931. Two days later, three more reached the summit. Holdsworth created a record for high-altitude skiing: he used his skis between camps IV and V at about 7231 m.

Nanda Devi

Hugh Ruttledge in 1926 and 1932, and the Shipton–Tilman team in 1934, had tried to find a suitable approach to Nanda Devi. The exploration and break-through into the Nanda Devi basin by Shipton and Tilman and their Sherpa companions in 1934, was not only the culmination of half-a-century of attempts to do so, but it was also one of the finest exploits of Himalayan exploration and adventure. And it was done on a shoestring budget, thus showing what could be achieved without heavy expenditure and large expeditions. Finally in 1936 a compact eight-man team of four Americans and four British under the leadership of T. Graham Brown determined to force their way up to the peak. Their porters deserted them, but the team had six very experienced and reliable Sherpas; after considerable difficulty they were able to reach the Nanda Devi sanctuary. On the mountain the Sherpa strength dwindled further, due to sickness. Nevertheless, on 29 August, from the bivouac at 7385 m, Tilman and Odell were able to reach the peak of Nanda Devi (7891 m) after a nine-hour climb. This excellent performance was due to the very good teamwork of all the eight members of whom T. Graham Brown, N.E. Odell, H.W. Tilman and Peter Lloyd were British, and W.E. Loomies, Charles Houston, Arthur Emmons and Adams Carter were

American.

Other expeditions in the latter half of the thirties were by a Japanese team which climbed Nanda Kot (6926 m) in 1936, their first in the Himalaya; and Spencer Chapman's and Pasang Dawa Lama's ascent of Chomolhari (7385 m) in Bhutan. In Sikkim, C.R. Cooke made the first ascent of Kabru, a 7385 m high mountain.

AFTER THE SECOND WORLD WAR

The Second World War had put a stop to all serious climbing in the Himalaya. Young Indian climbers, however, made their debut on the mountaineering scene in 1942, when three fifteen-year-old boys of the Doon School led by their British masters, Holdsworth and Martyn, set up camp at a height of about 5846 m on one of the glaciers above Badrinath.

The ascent of the first 'eight thousander' (more than 8000 m), Annapurna, by a French expedition led by Maurice Herzog in 1950 was the most spectacular and talked-about mountaineering event in the next few years. During their descent from the peak, the summit party comprising Maurice Herzog and Louis Lachenal, and their support team, were overtaken by one calamity after another. Frost-bite, snow blindness, falling into crevasses, being buried in soft snow, and avalanches, were the kind of ordeals the party survived. The credit for the success of this epoch-making achievement must go to every member of the team and the accompanying Sherpas. Nonetheless, Herzog's determination and 'do or die' spirit played a vital part in the final outcome.

The same year (1950), the Norwegians climbed the 7700 m Tirich Mir in the Hindu Kush range.

The Conquest of Everest

While in the past the Germans had concentrated on Kanchenjunga and Nanga Parbat, the British had jealously kept the 'Conquest of the Third Pole' (Everest) as their aim. However, during the six years of war, the main contenders in the battle of the peaks were too preoccupied killing one another to raise their eyes towards the sublime heights, and the Goddess Chomolungma remained in her splendid isolation for more than a decade.

After the war, the first thrust into the Everest region from the Nepal side was made by Charles Houston and Tilman in 1950. This was a very small expedition. After establishing their firm base in Thyangboche, the two climbers and four Sherpas, with provisions for six days, left the monastery on 16 November. Their reconnaissance of a possible route to the summit was mainly a visual one from the eastern slope of Pumori. Tilman's report only confirmed the pessimism expressed by Mallory thirty years earlier on this approach to Everest.

Despite Tilman's discouraging report, the British refused to give up. In 1951, a strong reconnaissance party set out under Eric Shipton. From the Pumori face they were thrilled to find that after the icefall, at least up to the South Col (8000 m), the going did not appear to be as impracticable as feared earlier; but they could not observe the Summit Ridge from there. The immediate obstacle was the Khumbu icefall and the first three attempts to negotiate it were unsuccessful. Ultimately they reached the top of the icefall on 28 October but the party did not venture beyond. To quote W.H. Murray, 'Our ascent therefore was particularly useful not only in proving the icefall climbable, but in breaking a psychological barrier and showing that the sense of heavy menace was illusory.' Shipton submitted the report of his successful reconnaissance on his return to London, and preparations for the next expedition were started at once.

The Swiss Foundation for Alpine Research had already obtained permission to climb Everest both in the spring and the autumn of 1952 which made the British somewhat unhappy. The Swiss suggested a joint Anglo-Swiss expedition and two members of the Swiss Foundation went to London to negotiate on the leadership issue, but no settlement was reached. The British abandoned their plan for 1952. For the Swiss, it was their first acquaintance with Everest, and for most of the members of the expedition their first experience of the Himalaya. They tackled one obstacle after another with skill and dogged determination, with exemplary team spirit and uncanny mountain sense. The star performers were Raymond Lambert and Tenzing Norgay. On 27 May 1952 they reached their final camp at a height of nearly 8400 m. Lambert describes the experience thus:

A dreadful night began. Nothing as it were to eat, a little cheese, a sausage, a candle to melt a little snow and that was all. In that little tent clinging to the ridge of Everest there commenced a terrible struggle against the cold. We thumped each other throughout the night rubbing our limbs which were gradually going numb.

If there had been some previous experience and better logistical support, the spearhead of the expedition would have been adequately fortified for the task ahead. Next morning Lambert and Tenzing started early. Preparation took very little time as they had slept with all their clothes on and had nothing to eat or drink. In their state of exhaustion and dehydration, after five-and-a-half hours' struggle they had climbed only 200 m and at 11.30 a.m. they decided to return. The post-monsoon expedition fared worse. In the middle of November, the extreme cold and strong winds forced climbers to retreat from a height of 8600 m.

Nevertheless, the Lambert–Tenzing team's performance in the spring was remarkable. Given better oxygen equipment and more favourable weather conditions, they might have climbed Everest. However, they did show the way and of this they are justifiably proud—even more than of their success in reaching the summit in 1956.

At 11.30 a.m. on 29 May 1953, two men for the first time stood on the highest point on earth. The two men were Edmund Hillary, a New Zealander, and the same Tenzing who a year earlier had reached a height of 8600 m with the Swiss by the same route. It was fitting that a British expedition was the first to be crowned with success on this mountain. From 1921 to 1938 they had returned again and again, suffering set-backs and casualties, but never giving up, never accepting defeat.

Mountaineers everywhere applauded it as a great event. India, not yet a mountaineering nation, gave a hero's welcome to the team. Special gold medals were struck to be awarded to the leader, John (now Lord) Hunt, and to the two climbers. The British achievement coincided with the coronation of Queen Elizabeth, and the British understandably celebrated it as a great victory and an auspicious start to the second Elizabethan era.

BIRTH OF INDIAN MOUNTAINEERING

Success on Everest was an event of great significance for India. To commemorate Tenzing's great achievement, Jawaharlal Nehru and Dr B.C. Roy decided to set up a mountaineering school in Tenzing's home town, Darjiling. Thus in 1954, with the late Major Nandu Jayal as its first Principal, Tenzing as the Director of Field Training, and a band of half-a-dozen veteran Sherpas as instructors, the Himalayan Mountaineering Institute came into existence. It was organized on a plan prepared by the Swiss Foundation for Alpine Research with the help of the Mountaineering School at Rosenlui.

With his youthful spirits and consuming passion for mountains and mountaineering, Jayal not only organized carefully planned training programmes, but had the Institute functioning in a very short time. Tenzing trained the youth of the country in mountaineering and inspired in them a spirit of adventure. In the first ten years, in addition to a number of basic courses, the Institute's advanced courses had students climb high Himalayan peaks like Kamet, Abi Gamin and Sekang, and attempt Saser Kangri and Nanda Devi. In 1958, barely four years after the Institute started, an Indian team successfully climbed Cho Oyu in Nepal, the seventh-highest mountain in the world. Tragically, Nandu Jayal lost his life in this expedition.

In 1956 the Swiss Foundation for Alpine Research organized a strong team composed mainly of Bernese climbers under the able leadership of Albert Eggler. Without much difficulty, two climbers made the first ascent of Lhotse, the fourth highest mountain, on 18 May. Then on two successive days, 23 and 24 May, two climbers each climbed Everest. True mountaineers, the modest Swiss came to the Himalaya, achieved all this, and returned to their native land without any fanfare or publicity. Not many knew about these achievements until much later.

Nanga Parbat (8126 m)

Five weeks after the first ascent of Everest, Nanga Parbat, the 'Killer Mountain', which had already taken thirty-one lives was climbed by a lone mountaineer, Herman Buhl. This Austro-German expedition was led by Dr Herrligkofer, the stepbrother of Willy Merkl. They established five camps above base camp. Herman Buhl and Kempter were to make the first attempt from camp V (approximately 7000 m). There was a storm at night; at 1 a.m. Buhl woke up Kempter, and left alone at 2.30 a.m. expecting Kempter to follow. Kempter tried to catch up with Buhl but could not and after reaching a height of 7692 m he returned.

With frequent rests, Buhl kept plodding on; he was three hours behind his schedule. Climbing with great caution and using the last reserves of his strength and endurance, he crawled the last lap on all fours, and reached the summit just before 7 p.m., after sixteen hours of climbing. During the return journey Buhl had to bivouac on the exposed ridge, standing the whole night above 8000 m. At 4 a.m. he started his descent, a great struggle in his dehydrated and weak state, when he was met by Ertl who helped him to reach safety. Even though Buhl's going alone was a subject of much criticism, his remarkable determination and endurance created a deep impression on the mountaineering community.

K2 (8611 m)

In 1953, fourteen years after his first attempt, Charles Houston brought a young team to his old enemy, K2. The expedition established eight camps, the highest being at a height of 7846 m. It was planned to set up camp IX for the final bid. By 2 August eight climbers, all well acclimatized, strong and confident, had assembled in camp VIII. Logistically they were well placed; they had rations and fuel for ten days. All they needed was three or four days of good weather, but their luck was out. The weather turned inclement the very first night, and a blizzard raged for five days; in the meanwhile Gilkey fell very ill and collapsed with thrombophlebitis. Priority was given to bringing the stricken climber down to safety without loss of time.

On 10 August on the way down, one of the frostbitten members fell and nearly dragged the whole party down with him. Fortunately, Schoening was able to hold the five climbers of the two ropes on his belay. While this rescue drama was on, Gilkey who had earlier been lowered down the steep ridge, was swept away by an avalanche. Thus ended Houston's second attempt. It was a miracle that the whole party did not meet Gilkey's fate.

K2 was finally climbed in 1954 by a strong Italian team under the leadership of Ardito Desio, Professor of Geology at Milan University. Like their predecessors, this expedition was also dogged by atrocious weather which lasted for more than five weeks. The Italian summiters Achille Compagnoni and Lino Lacedelli completed the ascent despite the fact that their oxygen equipment failed nearly 246 m below the summit. On the return journey, in darkness, they encountered a small avalanche, had a twenty-metre fall from an ice wall, and faced extreme wind-chill conditions, but were able to come through without too much damage to themselves.

Kanchenjunga (8598 m)

'There is no doubt', wrote John Hunt, 'that those who first climb Kanchenjunga will achieve the greatest feat of mountaineering, for it is a mountain which combines in its defences not only severe handicaps of wind, weather and very high altitude, but technical climbing problems and objective dangers of an order even higher than we found on Everest.' The British team which climbed Kanchenjunga in 1955 from Nepal was led by Dr Charles Evans who in 1953 was a member of John Hunt's successful Everest expedition. With Tom Bourdillon, he was on the first rope that preceded Tenzing and Hillary.

Two teams of two men each reached a point some metres below the summit. The climbers stopped short of the highest point, in deference to the religious sentiments of the Sikkimese, who consider Kanchenjunga a sacred mountain whose summit must not be desecrated by human feet.

Makalu (8481 m)

In 1954, the French explored the Makalu region during which they climbed a nearby mountain Chomo Lonzo (7797 m). The next year Jean Franco planned his expedition to Makalu meticulously and good teamwork resulted in all nine climbing members reaching the summit on three successive days, 15, 16 and 17 May. The team's performance on Makalu was a great achievement but compared to Herzog's Annapurna story which attracted worldwide attention, it lacked drama because everything went very smoothly. 'Expeditions in which nothing goes wrong', remarked Franco, 'are as lacking in history as nations which are happy.'

Cho Oyu (8153 m)

Cho Oyu which was first climbed in 1954 by Herbert Tichy, an Austrian, with Pasang Dawa Lama, was the Indian maiden attempt at a major peak. The Indian team, composed of fairly inexperienced mountaineers, was led by a Bombay solicitor, K.F. Bunshah. While leaving to join the expedition after handing over as Principal of the Darjeeling Institute, Nandu Jayal had remarked, 'After climbing Cho Oyu, my next destination is Everest.' But that was not to be. Cho Oyu was climbed by Sonam Gyatso and Sherpa Pasang Dawa Lama on 15 May 1958, but Nandu Jayal who died of pulmonary oedema, had to be left behind.

Though marred by tragedy, this success spurred the Indian Sponsoring Committee to book Everest for 1960. Only the loftiest mountain seemed to be a worthwhile goal for the Indians now.

Indians on Everest

By the time the necessary decisions were taken, adequate funds provided, and the leader selected, it was already August 1959. Within a period of five or six months, manufacture and procurement of clothing, equipment and other stores were arranged, including the import of oxygen cylinders and masks; prototypes had to be made for some of the items, and their field trials held before manufacture. After the selection of the team a pre-Everest training course was held which gave an opportunity to members of the team to go through their paces and get to know one another. With the leader's organizational capacity and the fullest co-operation of everyone, the task was completed on schedule. The expedition was able to set up a summit camp at 8492 m. At 3 a.m. on 25 May 1960, Sonam Gyatso, Nawang Gombu and Narinder Kumar got out of their sleeping bags and some hours later were on their way to the summit. The weather was none too good and was deteriorating fast. Strong winds and powder snow made visibility poor and the going very tough. It took them many hours to reach the foot of South Summit (8754 m). Further progress became very slow and extremely hazardous, and they reluctantly retreated when they were less than 215 m from their goal. Soon the monsoon struck with all its fury, and the expedition had to withdraw hoping for another date with the mountain.

With two special expeditions in 1961 (to the virgin peaks of Annapurna III and Nilkanth), both successful, the second Indian expedition in 1962 had a more experienced team and Major John Dias was selected to be its leader. The summit team consisting of Kohli, Sonam Gyatso and Hari Dang, reached a height of 8846 m before they were beaten back by impossible weather conditions. The team, however, set up some records in human endurance. The summit team spent three consecutive nights at the last camp at 8615 m, more than half the time without oxygen, having spent the preceding two nights at South Col; Gurdial Singh and one Sherpa were at South Col (8000 m) for six days, most of the time without oxygen. The team's doctor, Captain M.A. Soares, spent forty-five consecutive days at camp III (6523 m).

Other Expeditions to Everest

In 1963 Norman G. Dyhrenfurth led a very strong American team to Everest. On 1 May, James Whittaker and Nawang Gombu (Tenzing's nephew) reached the summit. For the next two weeks the weather led all to a standstill.

For the subsequent effort, Dyhrenfurth had planned a two-pronged simultaneous ascent of Everest. Finally on 22 May, Barry Bishop and Luke Jerstad climbed by the conventional South-East Ridge route and reached the summit at 3.30 p.m. William Unsoeld and Thomas Hornbein went by the new West Ridge route and reached the top after a tiring and difficult ascent, but too late, at 6.15 p.m., to meet the first party on the summit as planned. They came down by the South-East Ridge route, thus completing the first high traverse of Everest. At the foot of the South Summit, Barry Bishop and Jerstad were waiting for the West Ridge team. By the time Unsoeld and Hornbein came down, it was completely dark but fortunately they were able to make contact with the waiting pair. All the four members were forced to bivouac in the open at a height of 9500 m. The weather gods were kind, there was no strong wind at night and the party escaped without too much damage. However, all except Hornbein suffered from severe frostbite.

Meanwhile the Indians were preparing for their third attempt in 1965. They organized expeditions to Trisul, Nanda Devi, and Panchuli in 1964, of which the last two were successful. Twenty-six climbers were selected for a pre-Everest expedition to Rathong Peak in western Sikkim. M.S. Kohli had already been chosen as the leader with Major Narinder Kumar as his deputy. A team of fifteen was selected on 14 November 1964. This team left for Everest on 25 February 1965 and made the first successful ascent 85 days later. In between the expedition had to face several setbacks, the worst of all, as usual, on account of the weather. After an early start and encouraging progress on the mountain, bad weather forced the party back to the Base Camp. 'There had not been a single day from 25 April to 11 May when the wind's roar could not be heard at the Base Camp', wrote Kohli in one of his despatches. When the weather improved Kohli again moved up his mountaineers. On 20 May at 9.30 a.m. the first summit team of A.S. Cheema and Nawang Gombu reached the summit. That made Gombu the only man in the world to have made two successful ascents of Everest. Sonam Gyatso and Sonam Wangyal made the second ascent on 22 May. Two days later the hat-trick was performed by C.P. Vohra and Ang Kami who were the third pair on the summit. On 29 May, the twelfth anniversary of the first attempt on Everest, the fourth team of three climbers, H.P.S. Ahluwalia, N.C.S. Rawat and Assistant Sardar Phu Dorji reached the coveted peak. Never before had three men stood on the summit together, nor had four parties successively reached the peak. The record of nine men on top remained unbroken until the Franco-German expedition of autumn 1979. To Indian mountaineering it gave a tremendous boost. From one or two expeditions a year to peaks of 6000 m and above, within five years the number became fourteen or fifteen and in the mid-seventies it rose to thirty-five or so a year. Many difficult peaks have been climbed by the Indians, including Hathi Parbat, Shivling, Changabang, Sakang and Sasar Kangri.

Development of Advanced Techniques

1965 became a significant milestone in the history of mountaineering in the Himalaya. A century earlier climbers had made the first ascent of Matterhorn and thereafter Alpinists turned their attention from conventional ascents to more difficult routes, first of ridges and then of Great Walls. The spirit of man found new and more difficult ways of getting to the top and used increasingly sophisticated equipment. Even so, progress was painfully slow: it took almost 66 years for the Matterhorn to be climbed by the hazardous North Face. The techniques and equipment which had been perfected in the Alps were now brought into play in the Himalaya.

From 1965, Nepal had remained closed to foreign expeditions for four years. When the ban was lifted in 1969, it became possible for experienced climbers to telescope developments in the Himalaya into the seventeen years from the first ascent of Everest, to the attack of equivalents of the Eiger North Wall and the North Face of the Matterhorn, at Himalayan altitudes. The British were the first to make a bid on the South Face of Annapurna, a formidable mountain wall with 4000 m of steep rock and ice leading up to its 8000 m summit.

Attempts on Everest South-West Face

A new generation of mountaineers now started looking for more difficult routes up major mountains that had already been climbed. Perhaps the most challenging and certainly the most glamorous prize was the South-West Face of Everest. In the short period from 1969 to 1978 after the Nepalese Government relaxed its restrictions on entry, five expeditions tried the Face and failed, defeated by a combination of strong winds, intense cold, and problems of high altitude, in addition to severe climbing conditions.

The Japanese made a light reconnaissance in the spring of 1969 followed by a reconnaissance in force in autumn the same year. They were able to establish camp V at 6500 m before winding up the exploratory plan. A massive Japanese attempt the following spring failed because of a couple of mishaps in which the team lost six Sherpas, and considerable time, thus delaying the build-up.

Since 1970 there have been two expeditions from the Nepal side every year from several countries. The Japanese reserved Everest in 1970 for both spring and autumn. Four climbers gained the summit in the first expedition, including the famous Naomi Uemura in the first party. In the second expedition Y Miura skied down from the South Col (7880 m) to the Western cwm using a parachute to break his speed on the precipitous slopes. A film of this daring exploit has thrilled audiences throughout the world.

In 1971 Norman Dyhrenfurth had organized a multinational group of individually strong climbers representing six countries. Even the lure of

Everest failed to bind this motley collection of mountaineers into a well-knit team, and the expedition broke up. India paid a heavy price in this experiment in international mountaineering. Major Harsha Bahuguna, a fine gentleman and mountaineer, died on this expedition under tragic circumstances. Even after the virtual break-up of the expedition some members attempted to force their way up the South-West Face, but failed for want of adequate support.

In the latter half of 1971, a large Argentinian expedition failed, as did Dr Herrligkoffer's European team in the spring of 1972 and Chris Bonington's team in the autumn of the same year. In 1973, a Japanese expedition placed four men on top, while a rather extravagantly organized Italian Army group succeeded in sending six members and two Sherpas to the summit.

Chris Bonington returned to the South-West Face in the autumn of 1975. The team, consisting of eighteen strong climbers supported by sixty high-altitude Sherpas, was equipped with protective clothing and special iron-mongery to withstand the extremely severe climatic conditions. Doug Scott and Dougal Haston reached the summit on 24 September, to be followed two days later by Boardman and Sherpa Pertemba. This was actually the first occasion on which the summit of Everest was reached by British mountaineers. The achievement was marred by the disappearance of Mick Burke, the BBC cameraman who climbed the last stretch alone, and failed to return. Both parties climbed the last few hundred metres from the South Summit onwards along the South-East Ridge. The true ascent has still to be made the whole way up the South-West Face.

Women on Everest

In 1975, as many as fifteen climbers from three expeditions reached the summit. It was also the International Women's Year and so, fittingly, the Goddess Chomolungma welcomed a Japanese housewife in the company of Sherpa Ang Tsering on the top of the world. Later, another woman, this time a Tibetan, made it to the top along with eight other climbers in a Chinese expedition. It was the first ascent of the mountain from the North, recalling the heroic attempts of the British in the years before the war.

In 1976, two climbers of a British team reached the summit. Both of them were badly frost-bitten, and five men lost their lives on this fateful venture. The American Bicentenary Expedition also completed its mission in the post-monsoon period of 1976, with two climbers hoisting the American Stars and Stripes on the top of Everest.

Everest without Oxygen

'Will Everest ever be climbed without oxygen?' Climbers had reached 8615 m without oxygen, and many people considered this the limit of human endurance, but Messner and Habeler proved otherwise in 1978. These intrepid Alpinists braved the physiological 'Death Zone' and reached the highest peak on earth without oxygen. In the post-monsoon season of 1978, three members of a joint European expedition, one German and two Sherpas, also made the summit of Everest without oxygen. If there were any doubts about the first claim by Messner and Habeler, they have been put to rest by the repeat performance.

From 1921 to 1978 there have been forty-five attempts at scaling Everest. In fifteen successful expeditions eighty-one climbers (five men without oxygen) including three women have stepped on the coveted pinnacle: of these, twenty-five did so in 1978! But these achievements have not been possible without much loss of life. Over forty-eight climbers have lost their lives in enabling others to attain the highest point on earth.

Changabang (6864 m)

The development of modern techniques employing the latest artificial aids, encouraged mountaineers to take on technically more demanding peaks in the Himalaya. The South Face of Annapurna and the South-West Face of Everest entailed going the harder way up major peaks already climbed. Some mountaineers, however, started looking for smaller mountains of between 6000 and 8000 m which had no obvious or easy way to the top. The first objective was the 6864 m high Changabang which was described by Longstaff as 'the most superbly beautiful mountain I have ever seen, its North-West Face a sheer precipice of 5000 feet'.

An Indo-British expedition jointly led by Chris Bonington and Balwant Sandhu climbed this difficult peak in the spring of 1974. Six climbers, four British and two Indians, reached the summit.

The success of the Indo-British expedition focussed the attention of other Alpinists on this challenge and in four years as many expeditions reached its summit by different routes. The three expeditions that climbed in 1976 alone, were the Japanese expeditions by the South Face in June 1976; the two-man British team by the West Face in October 1976, and in the same month another British expedition by the East Ridge. Finally an Anglo-Polish party succeeded on the extremely difficult South buttress in the autumn of 1978.

Indians on Kanchenjunga

In 1977 an Indian Army team led by Colonel Narinder Kumar climbed Kanchenjunga for the first time by the more difficult and hazardous South-East Ridge. They followed Paul Bauer's route from the Zemu glacier. Major Prem Chand and Naik N.D. Sherpa were the summiters. Like Charles Evans' team earlier, they also left the last few metres untrodden, honouring the sentiments of the Sikkimese who hold Kanchenjunga sacred. Paul Bauer paid high tribute to this climb. Dr Nishibori, President of the Japanese Alpine Club considered it 'one of the greatest achievements in mountaineering history.'

The Tiger of the Snows

It has been said that if God had not created the Sherpa, Everest might be yet unclimbed. 'The man from the East' which is what Sherpa stands for, is so inextricably a part of mountaineering in the Himalaya that there is hardly any incident or adventure in the last fifty years which has not included him. Like a strong mountain goat, humble, sturdy and swift, he climbs, ferrying loads up the precipitous and treacherous ice, rock and snow slopes, through inclement weather, biting winds and many other serious mountain hazards until the leader of the expedition prevents him. The fantastic endurance and equanimity of the Sherpa, his friendship that never fails, company that always cheers, his assurance born of skill and courage, have few parallels in mountaineering history. He is the true Tiger of the Snows, and without him so much by so many could not have been achieved, certainly not at such little cost in money and life.

MOUNTAIN ENVIRONMENT

Considering the magnitude of the Himalaya, the impact of the increasing number of mountaineering expeditions on the environment might not appear to be of too great consequence. Nevertheless it is also not insignificant. The Nanda Devi sanctuary, which has only one entrance, has been severely affected by more than two dozen expeditions a year in recent times. With the opening of fresh trails, hunters, poachers, trappers and other exploiters of forest wealth have followed and as a result, there has been unchecked and indiscriminate destruction of flora and fauna. The expeditions themselves have been guilty of denuding the higher slopes of juniper, rhododendrons and other bushes for fuel. With the disappearance of trees and other vegetation, the rate of erosion and landslides has increased, exposing bare rocks on which nothing can grow. It is not surprising, therefore, that large areas and innumerable slopes on popular expedition routes, and the grounds frequently used for base camps, are being converted into barren patches.

Man's thoughtlessness has adversely affected the unspoilt beauty of places like Kishtwar, Solu Khumbu, Nanda Devi Sanctuary, Manali and many sites in hitherto unpolluted places in the high mountains. It has become fashionable to visit places like Thyangboche, Khumbu glacier or the base camps of other major mountains, leaving behind, not only wanton destruction but trails of garbage.

In the decade ending 1974, the traffic to Solu Khumbu increased from 300 a year to 3400—an example of the new human invasion of the higher

Himalayan areas. Gone are the days when a few surveyors quietly roamed
about, and when there were less than a dozen expeditions in the whole
Himalaya. The present is an era of organized recreation with hundreds of
mountaineers, trekkers and tourists visiting the Himalayan region each year.
Their presence makes heavy demands on scarce resources such as fuel, water
and space in these limited areas. The destruction of vegetation here is
almost irreversible since at elevations above 2500 m regeneration is difficult.
This influx has unfortunately now spread to Ladakh, which for centuries
had maintained a delicate ecological balance between limited numbers of
human beings and very limited natural resources.

CHAPTER 12

POPULATION AND SOCIETY IN THE HIMALAYA

B.K. ROY BURMAN

The Himalaya is the youngest mountain range of the world, and attained its full growth only in the lower Pleistocene period. It is suspected to be the primeval cradle of mankind; the Siwalik system, starting from near Dehra Dun, contains fifteen genera of anthropoid apes, the highest mammals in the then world, some of which are believed to form links in the line of human ancestors (Wadia, 1965, p. 146). There are in any case prehistoric relics which suggest the presence of man in the western Himalaya during the middle Pleistocene period. This age is marked by three glacial advances and retreats. In the boulder conglomerate of the second glacial period, split pebbles of quartzite are found on the bed of the river Sohan. As opinion is divided on whether these are the results of man's deliberate action, they are described as pre-Sohan specimens. From the close of the second ice age, chopper-chopping and flake and blade industries began to occur on the foot-hills of the south-west Himalaya. Thus even at the beginning of human history, two stone-making traditions seem to have existed in the area, but so far no remains of the man who made these tools have been found. It is therefore difficult to say whether these different traditions belong to two racial groups or not. There is however evidence to show that some sort of fusion, either of races or of tool traditions, took place in the later phase of the stone age (Sankhalia, 1973, p. 38). Here it is interesting to note that floristically also the Himalayan belt has been a meeting point of Euro-Mediterranean and Western Chinese affinities. Further, though it has served as a bridge, it has also been a barrier, excluding much of the Tibeto-Chinese component and facilitating both derelict and endemic elements (Seth, 1978, p. 9). In contrast to the western Himalaya, the eastern Himalaya is a relatively unexplored area as regards pre-historic and proto-historic periods (Sankhalia, 1973, p. 38).

The Himalaya can be divided into two regions—the western Himalaya, reaching up to the eastern border of Nepal; with the eastern Himalaya covering Sikkim, Darjiling in West Bengal and Arunachal. As a lateral extension of the eastern Himalaya, the hill ranges of Nagaland and Meghalaya will also be discussed here.

The western Himalaya has a greater representation of conifers. There are a few primeval forests left, but scrub and grass lands have invaded erstwhile forest territory, wherever it has not been put to organized human use for agriculture and settlement. This region can be divided into a number of phyto-geographical sectors as follows: north-west sector (Karakoram in the north to Ladakh); Punjab sector (Kangra to Bushahr, west of Satlaj); east sector (Tehri Garhwal to Kumaun in India and western Nepal up to Karnali-Gandaki). In the north-west sector the vegetation is semi-desert, with a few *juniperous macropoda* reaching tree dimensions, with poplars and willows along water ways. Here alpine meadows are characteristic with herbaceous elements predominating. In the Punjab sector, commercially valuable deodar and *chir* forests along with blue pine are the most important, but oaks are also widespread. In the eastern Kumaun forests deodar gradually becomes infrequent and hemlock (Tsuga) appears in the extreme end (Seth, 1978).

The eastern Himalaya subregion with a lower altitude, higher snowlines and higher precipitation has its own botanical identity. Conifers are far less important although some pines of east Asian affinity and fir and birch occur; oaks and chestnuts are characteristic of the temperate zone. Sub-Alpine birch and fir mixed with junipers, rhododendrons and hemlock occupy the drier ridges (Seth, 1978).

This variegated landscape has on the one hand contributed to the making of man and human institutions, and, on the other, bears the mark of man's capacity to make and unmake his environment. The Himalaya and humanity have been confronting and negotiating with one another through the ages.

LAND AND MAN IN THE HIMALAYA

As already noted there are evidences of the presence of man in the western Himalaya, even at the dawn of human life on earth; the eastern Himalaya seems to have been populated much later. Today one comes across human settlements almost all over the Himalaya, but the density of population per unit of land is much less than that in the plains of India.

The density of population varies from 1.09 per km in Ladakh to 274.70 per km in Srinagar, both in the western Himalaya. If the eastern Himalaya is considered separately it varies from 2.57 per km in Lohit to 254.23 per km in Darjiling.

Density of population figures for the total area do not however represent the density of population for the actually habitable or cultivable area. For instance, in the sector falling within Uttar Pradesh, the higher ranges supporting scanty to little forest growth account for more than one fifth of the area. While considering population density such areas as well as ravines, deep gorges and waterways will have to be left out.

Another indicator of the level of techno-economic adaptation is the settlement pattern in diverse ecological settings.

Settlement Pattern

Bose (1972, p. 96–7) gives an admirable account of the distribution of rural and urban settlements in various altitudinal zones, which is summarised here:

1. 0 to 1000 metres: There are deep gorges and sharp water partings, and the area is covered with tropical forests infested with wild animals. The climate is also warm and humid. There are few settlements of any importance.

2. 1000 to 2500 metres: This may be called the optimum belt in the Himalaya and the majority of people live in this belt. Here the climate is cool and the river terraces well developed.

3. 2500 to 3000 metres: Here the river gorges are deep and one often comes across U-shaped cliffs and thundering waterfalls. The climate is cool and the land covered with coniferous forests. There are only some villages with cultivation in favourable places.

4. 3000 to 5000 metres: There are alpine meadows and much grazing. One frequently comes across semi-nomadic shepherds and traders. There are also some temporary summer dwellings.

5. 4000 to 5000 metres: This belt is cold, rocky and snowy. There are some pastures on favourable spots which semi-nomads visit now and then.

6. Above 5000 metres: There are rugged empty regions of rock, ice and glaciers.

This account brings out the capacity of the people of the Himalayan highlands to take advantage of the endowments of nature to a remarkable extent. But it also suggests that man continues to depend on nature more at the primary level of extraction and processing than at the secondary level of processing. This point will become evident when distribution of population in rural and urban areas, urban density and participation of workers in diverse occupations and industrial categories are considered.

Density of Population in the Urban Areas

Density of population in the urban areas gives a better basis of comparison than when total areas are considered, as it can be assumed that within urban areas there will not be many uninhabitable places. There is a wide range of variation of urban density. As against the all-India urban density of 2504.45 per km^2, it varies from 183.72 (in Garhwal district) to 4097.31 (in Srinagar district) in the western Himalaya, and 999.42 (in North Cachar Hills) to 4516.59 (in Darjiling district in the eastern Himalaya).

It would be interesting to examine to what extent the low urban density in Garhwal and the North Cachar Hills district reflect constraints of physiography, and to what extent failure of mobilization of the productive forces which exist within reach of the towns. Similarly, the high urban densities in Srinagar and Darjiling districts have to be examined in terms of the interplay of endogenous and exogenous forces, such as the endogenous growth of

productive forces, on the one hand, and the fleeting attraction of tourist
interests on the other. Systematic studies in this matter are not known to
have been done.

Distribution of Population in Rural and Urban Areas

Population distribution in the rural and urban areas of the states and Union
Territories, the bulk of whose areas lie in the Himalaya region, is as follows:

	1971 Census	
	Rural	*Urban*
Jammu & Kashmir	81.41	18.59
Himachal Pradesh	93.01	6.99
Sikkim	90.27	9.23
Arunachal	96.30	3.70
Meghalaya	85.45	14.55
Nagaland	90.05	9.95
India	80.09	19.91

The statement shows that except for Jammu and Kashmir and to a certain
extent Meghalaya, in all the states and Union Territories of the Himalaya
region, urbanization has been much below the national average. In Jammu
and Kashmir and Meghalaya again, urban growth possibly reflects the
functioning of extraneous factors like tourism or location of administrative
offices more than the development of endogenous productive forces.

A clear picture emerges when the distribution of the rural and urban
population is examined at the district level. A synoptic view of the distribution of districts with reference to the percentage group of urban population can be obtained from the following statement:

State/Union Territory	Total number of districts in the Hima-laya	Number of districts without urban popu-lation	Number of districts with percentage of urban population to total population				
			less than 5%	*5-9 %*	*10-14 %*	*15-19 %*	*20% & above*
Jammu & Kashmir	10	—	1	6	1	x	2
Himachal Pradesh	10	2	3	3	1	x	1
Uttar Pradesh	8	x	4	2	x	x	2
West Bengal	1	x	x	x	x	x	1
Sikkim	4	x	3	x	x	1	x
Assam	2	x	1	1	x	x	x
Arunachal	5	2	2	1	x	x	x
Nagaland	3	1	x	x	1	1	x
Meghalaya	2	x	1	x	x	x	1
Total	45	5	15	13	3	2	7

The statement shows that out of 45 districts in the Himalaya region, there are five without any urban population. Two of the latter are in Himachal Pradesh and Arunachal each and one is in Nagaland. At the other end there are seven districts with more than 20 per cent urban population in each.. Interestingly enough, one of them—Simla—is in Himachal. In fact in Simla district 31.84 per cent of the population live in the urban area. There are two other districts in the Himalaya region, namely, Dehradun (47.08 per cent) and Srinagar (51.14 per cent) where more than 30 per cent of the population are town-dwellers. These figures bring out not only the extreme disparities of development in different regions of the Himalaya, but also the discontinuities of development.

Further confirmation of this is obtained when the distribution of the towns by size-classes of population living in them is considered. With the extent of urbanization at the district level, there is no continuum in the distribution of individual urban settlements either. In Jammu and Kashmir particularly, two cities seem to tower over the rest of the region. In Arunachal and Nagaland, however, one can see new formations in the process of taking shape through the interplay of both internal and external factors.

Distribution of Villages by Size-Classes

Compared to India as a whole, small villages predominate in the Himalaya region, except in Jammu and Kashmir and in Darjiling district. Meghalaya and Nainital district, and to a certain extent Nagaland, however, provide a picture of a continuum from small to medium and large villages in terms of population.

The predominance of villages with small populations is obviously related to the constraints of the physical environment to a large extent. But the possible role of techno-economic factors, injected from outside, in determining the size of rural settlements should not be ignored. For instance in Jammu and Kashmir, Darjiling and Nainital, the emergence of big towns in the wake of tourism and the proliferation of administrative and servicing agencies has probably been an important factor for the presence of medium and large-sized villages.

The meagre urbanization and predominance of villages with small-size population may now be seen in the context of the pattern of participation of the population in the working force.

Participation Rates in the Working Force

Participation rates of males and females in the working force show that except for Jammu and Kashmir and Uttarakhand, in all the states and

Union Territories and subregions of the Himalaya region, participation rates are higher than the all-India average. The pattern of participation of males in the rural and urban areas of Nagaland however deserves special mention. Unlike in other states, here males participate in the working force at a much higher rate in the urban area than in the rural area. Again, the extremely low participation of females in the working force in Jammu and Kashmir and Uttarakhand, on the one hand, and the very high participation rates of females in Arunachal and Nagaland and fairly high participation of all these are clearer when the data are examined at the district level, and also when the distribution of workers by industrial categories is examined.

At the district level a very significant deviation from the state average is found in Ladakh where 29.24 per cent of the females are workers as against the state average of 3.90 per cent. In Himachal Pradesh the participation of females in the working force is found to vary from 11.72 per cent in Simla to 61.35 per cent in Lahul and Spiti as against the state average of 24.20 per cent. The only other district where the female participation rate is more than 50 per cent is Kinnaur (54.88 per cent). In Uttarakhand female participation rates represent a much wider spectrum, starting from 7.72 per cent in Naini-tal (followed by 8.52 per cent in Dehradun) to 61.53 per cent in Uttarkashi (followed by 60.59 per cent in Chamoli). In contrast in the whole of eastern Himalaya, if the Mikir Hills are left out, the spectrum is much narrower, starting with 21.61 per cent in Darjiling district and going up to 59.54 per cent in Subansiri district.

The inter-district differences in the subregions of western Himalaya and the broad pattern of difference between the western and eastern Himalaya represent cultural differences more than differences in physical environment. In the western Himalaya the districts with high male and female participation rates in the working force are the ones where the influence of Tibetan Buddhism is strong. Again, districts with very low participation of females in the working force are the ones where urbanization has been the highest.

The high participation rates of females in areas under the influence of Tibetan Buddhism in the western Himalaya and in the different districts of the eastern Himalaya do not appear to be related to the same factors. In areas influenced by Tibetan Buddhism in the western Himalaya, the high participation is linked up with the pursuit of traditional household industries, trade and commerce and cultivation under conditions of semi-serfdom. In the eastern Himalaya it is more related to the survival of communal ownership of resources which ensures the continuous employment of males and females, although with low economic returns. A study of the distribution of the workers by industrial categories will throw further light on this.

Distribution of Workers by Industrial Categories

Except for Darjiling district in West Bengal, cultivation is the predominant occupation everywhere. But whereas in Sikkim as many as 81 per cent of the workers are cultivators, in Darjiling district the corresponding figure is only 30 per cent, followed by 65 per cent in Jammu and Kashmir. But what is more significant is that whereas in India as a whole 43 per cent of workers are cultivators and 26 per cent are agricultural labourers, nowhere in the Himalaya region excepting Uttarakhand do agricultural labourers constitute more than 10 per cent of the workers. In Uttarakhand 20 per cent of workers are agricultural labourers, but in Arunachal and Nagaland only 2 per cent belong to this category. In fact unlike in many other parts of the country, in the Himalayan region most of the agricultural population consists of owner cultivators. (It is a sequel to the development activities that they are progressively losing their lands.) There is another side to the story as well: as against the all-India average of 3 per cent, in the Himalayan region 6–19 per cent of the workers are engaged in 'other services'. This is a mixed category, ranging from domestic servants under the bonded system or otherwise, to persons engaged in administrative, research and other services. The high percentage of 'other services' in some sections of the Himalaya seem to reflect the presence of bonded labour and village servants practically under the conditions of serfdom.

The low percentage of cultivators (30 per cent) of Darjiling district deserves special mention. It goes along with the fact that 26 per cent of the workers are engaged in livestock rearing, forestry, plantations, etc. In fact the majority of persons returned under this industrial category are employees in tea plantations. One wonders why, despite the presence of tea plantations with their potential of generating employment in several directions, landless agricultural labourers should exist in this district to such a large extent. One of the reasons seems to be that the profit generated in tea plantations is not ploughed back to develop linked productive activities; as a result, the multiplier effect of employment on plantation work is minimal. But there is also another important reason. In this district much cultivation is done on terraced land, which requires a very heavy input of labour and capital. This leads to the concentration of ownership of land on the one hand, and a growing army of landless labour on the other. The prevalence of landless agricultural labour in a high proportion in Uttarakhand might also be related to the same cause. Similar social problems, though in different forms, are found almost everywhere in the Himalaya where terraced cultivation is practised. The matter requires a systematic in-depth study.

Some attention should also be given to the presence of workers engaged in livestock and forestry at a percentage twice as high as the national average in Jammu and Kashmir. This is related to the functioning of pastoral semi-nomads in this state as well as in the adjoining state of Himachal, where as many as 3 per cent of the workers have been enumerated in the composite industrial category of livestock breeding, forestry, etc. The presence of the household industry sector at the same level as the national average in Jammu and Kashmir and at a slightly lower level in Himachal Pradesh is also a matter of some importance. What is however of great significance is that in these two states there has been some growth of manufacturing other than household industry. In the other states neither household industry nor non-household industry has been returned as the main occupation by any significant number of people, which does not however mean that household industry is not important in those states. In fact, all over the Himalayan region, household production of cotton and woollen textiles and bamboo and wood crafts continues quite vigorously. But in the pre-feudal social formations of the eastern Himalaya, they are generally practised as part-time occupations along with other productive activities and domestic chores. In the feudal setting of Himachal and the national, and international-market oriented production system of Jammu and Kashmir, these crafts have emerged as the specialized occupations of part-time craftsmen. Census data reflect differences in the social organization of crafts, rather than the magnitude of production and survival potential of the crafts.

The present set of census data provide another interesting sociological indicator which should not go unnoticed. At the national level, whereas 21 per cent of male workers are agricultural labourers, for females the corresponding figure is 50 per cent. This implies that women of the socio-economically weaker households participate in the work force more frequently. But in the rural areas of the Himalaya region, the proportion of women working as agricultural labourers is nowhere higher than that in the case of males except in Uttarakhand; at the same time, unlike in the country as a whole, the proportion of women workers as owner cultivators everywhere outnumbers the male workers of the same category. Uttarakhand is however an exception; perhaps the deviant pattern there has resulted from the large-scale out-migration of males belonging to agricultural labour households to the plains to eke out a living.

There are several possible explanations for the general pattern in the Himalaya: the most plausible one is that through the mechanism of sharp sex-wise division of labour, females have been able to hold their ground in the social organization of production in land-holding households. Hence they can participate in the working force as a way of life and not as a forced necessity. One may ask whether the specificity of the social organization of labour in the Himalaya is accounted for with reference to any aspect of Himalayan ecology, or whether one should seek for explanation elsewhere. A comparative study of the ethnographic literature for different parts of India shows that in the pre-feudal tribal communities sex-wise division of labour persists even now. In the areas where a feudal political economy prevailed, the general tendency had been to divest the women of their productive role and cover up this deprival by projecting an ornamental and mystic image in a non-productive sector. Where however feudalistic consolidation had combined with military adventurism, sex-wise division of labour could not be expunged to the disadvantage of females. In those areas the cognitive process tended to be blurred by a discordance between the desired image and the facts; while it was prestigious to project females as non-workers, as an empirical reality their contribution to the economy of the household was quite substantial.

With the spread of education and deep infiltration of market forces, the process of sex-wise non-recognition of labour is involuted. Male monopoly of the employment market is disrupted, and a new pattern emerges in the social organization of labour. The diversified pattern of participation of females in the working force in Meghalaya and Jammu and Kashmir suggests that this process is also taking place there. But in other areas the pre-feudal features persist, as among most of the tribal communities elsewhere in India. Hence rather than physical environment, it is political economy which seems to have played the more important role.

From this viewpoint, the future rests largely on the spread of education.

EDUCATION

Except for Jammu and Kashmir, Uttarakhand, Sikkim and Arunachal, other states and Union Territories and subregions in the Himalaya are more or less at par with the national average in the spread of literacy. But when sex-wise literacy figures are considered, female literacy in some states of Himalaya is found to be far ahead of the national average. For instance, even in the rural areas of Himachal Pradesh, Nagaland and Meghalaya the females are 25–50 per cent more literate than females all over the country. Again if the urban and rural areas are considered separately, urban literacy in Meghalaya, Nagaland, Darjiling and Himachal is considerably higher than the national average. What is however more significant is the fact that the difference in the levels of literacy between the rural and urban areas differs considerably in the various subregions of the Himalaya. Thus whereas rural literacy in both Nagaland and Meghalaya is around 23 per cent, urban literacy is 60.79 per cent in Nagaland and 65.22 per cent in Meghalaya. The corresponding figures at the all-India level are 23.74 per cent and 52.44 per cent respectively. Again whereas rural literacy is around 14 per cent in both Sikkim and Jammu and Kashmir, urban literacy in Sikkim is 46.17 per cent and in Jammu and Kashmir is only 38.17 per cent. With a rural literacy rate of 18.30 per cent, Uttarakhand has an urban literacy of 43.63 per cent; on the other hand Arunachal with a rural literacy of 9.79 per cent has an urban literacy of 50.46 per cent. It is obvious that these variant patterns cannot be explained only in terms of the physical environment of the different sub-regions of the Himalaya. It is equally obvious that urban literacy has not grown by drawing upon the rural base, or the reverse, urban literacy has not influenced the surrounding rural area in the same manner everywhere. In Jammu and Kashmir, Himachal, Uttarakhand and Darjiling district, there appears to exist some amount of continuity in the educational process of rural and urban areas; in contrast, in Sikkim and Arunachal particularly, the educational process is marked by sharp discontinuity.

When the sex-wise literacy differentials are examined, another interesting picture emerges. Compared to the national average, both male and female literacy is higher in Himachal Pradesh and Darjiling district; in no other area is male literacy higher than the national average. Female literacy is higher than the national average in Meghalaya, but male literacy is much below the national average. In Nagaland female literacy is more or less at par with the national average but male literacy is at a lower level. It is obvious that educational opportunities have not been used in a uniform manner by males and females all over the Himalaya. In addition to such factors as difference in access to or control of productive resources, socio-cultural (including religious) differences must have played an important role. In this matter it should be possible to gain additional insights if a comparative study of the progress of literacy in different districts of the Himalaya is made. Information about any such study is not readily available, but after a cursory glance at the data, a few brief observations are made here. In Jammu and Kashmir, the level of female literacy is higher in districts where Hindus are numerically predominant than ones where Buddhists or Muslims predominate. In Himachal as well, low female literacy seems to be partly related to the prevalence of Buddhism. In contrast, in parts of Arunachal, Buddhism has played some, though a very limited, part in the spread of literacy. In fact, the only tribe in the Himalaya that have an age-old script and written literature of their own are the Khampti of Lohit district in Arunachal, who follow the Hinayana form of Buddhism. In Meghalaya the higher rate of literacy among females and lower rate among males compared to the national average, is perhaps related to the matricentric social organization among the Garo and the Khasi. The spread of Christianity again has been an important factor affecting the level of literacy particularly in Nagaland, and partially in Meghalaya. In view of the importance of religion in several spheres of life in the Himalaya, a rapid appraisal of population distribution by religious affiliation would be useful.

DISTRIBUTION OF POPULATION BY RELIGION

A statement giving the percentage distribution of the population by religion up to the state level is furnished below.

The statement brings out the fact that except for Himachal and Uttarakhand, the pattern of population distribution by religion is entirely different from that in the rest of the country. Again, within the Himalaya region intersectoral differences are also quite considerable. While in India as a

State/Union Territory	Percentage of population professing to be							
	Hindu	Mus-lim	Chris-tian	Sikh	Bud-dhist	Jain	Others	Religion not stated
Jammu & Kashmir	30.42	66.85	0.16	2.29	1.26	0.02	N	N
Himachal	96.08	1.45	0.10	1.30	1.04	0.02	0.01	N
Uttarakhand (UP)	83.76	15.48	0.15	0.42	0.05	0.14	N	N
Sikkim	68.88	0.16	0.79	0.05	29.84	0.09	0.19	N
Arunachal	21.99	0.18	0.79	0.29	13.13	0.01	63.46	0.17
Nagaland	11.45	0.58	66.78	0.13	0.04	0.12	20.94	N
Meghalaya	18.50	2.60	46.96	0.12	0.19	0.05	31.45	0.13
India	82.72	11.21	2.80	1.89	0.70	0.47	0.40	0.01

whole, Muslims constitute 11.21 per cent of the population, they constitute 66.85 per cent in Jammu and Kashmir and 0.18 per cent in Arunachal. Similarly Christians constitute 2.60 per cent of the population at the national level, 0.10 per cent in Himachal, 46.98 per cent in Meghalaya and 66.76 per cent in Nagaland.

The process through which these diverse patterns of the incidence of various faiths have come into existence in the Himalaya is by itself an important aspect of the cultural dynamics of India. It may however be stated in a general way that except among migrants during the last one or two genera-tions, the various all-India or universal religions have been adopted by the indigenous population as alternative identities for facilitating their com-munication with the external world. Hence frequently the different religions exist as outer covers for the core faiths, which show a remarkable continuity with ancestral faiths. This is true particularly in the case of Hinduism, Christianity and Buddhism. But even Islam, in areas like Jammu and Kashmir, has a flavour of its own, different from that of the rest of the country. In Kargil tehsil, a sect called the Nur Buxia flourishes which seems to be an amalgam of Shia and Sunni denominations. In the Kashmir valley Muslim saints, called *baba rishis*, provide a spiritual umbrella to Muslims and Hindus alike. Again many Muslim communities not only retain Hindu gotra names, but also refrain from marrying within the same gotra.

The large percentage of people returning their religion as other than the religions of national importance in Arunachal, Nagaland and Meghalaya is a matter of great interest. These are indigenous faiths, mainly centering around natural phenomena, uncanny events and ancestor worship, which have been described by early ethnologists as animism or spirit worship. But this does not appear to be a correct appraisal of the religious mores. In the cognitive-affective process of people, belief in a supreme being is also uni-versal with varying degrees of involvement in human affairs; besides one occasionally finds an inkling of the sublime in them. Perhaps what marks out the indigenous faiths from the religions practised all over India and the world is the relative absence of systematization and formalization and relatively greater emphasis on the affective than on reflective aspects.

The presence of special features in religious beliefs and practices should not, however, suggest that the basic tenets of the different religions do not have any bearing on the social demography of the population concerned. For instance the low participation rate of females in the working force in Jammu and Kashmir is, along with other factors, linked up with the Muslim practice of purdah or seclusion of women.

MIGRATION

As already stated, the cultural drives and imperatives of social organization are different in the case of migrants. The 1971 census data on migration are not yet available at various levels; the 1961 census data show that the extent of in-migration varies from 1.30 per cent in Baramula to 48.78 per cent in Nainital in the western Himalaya and from 3.24 per cent in Tuensang to 27.10 per cent in Darjiling district in the eastern Himalaya. Broadly, these figures represent two different processes. In some areas, Nainital for instance, in-migration of a large amount of the agricultural population has taken place as a result of clearing vast tracts of land in the *tarai* for agricultural purposes. In a district like Dehradun again, in-migration is more related to urban growth as a result of the expansion of administrative agencies, educational facilities and other servicing organizations.

Some districts like Garhwal and Tehri Garhwal are known for large-scale outmigration of population, side by side with a small amount of in-migration. A good number of the migrants serve in the army, but a sizeable proportion also move on to Delhi, Lucknow, Bareilly and other cities in the plains and take up menial and semi-menial jobs. It seems that recently there has been some decline in the volume of migration of this order. But a consolidated picture of migration from the different districts of Himalaya to places outside their respective states is not available from the Census. In the eastern Himalaya as well, a tendency to move from the hill-tops to the foot-hills is noticeable. This is almost invariably related to the expansion of settled cultivation by clearing the forests. In several cases it has led to inter-state tension, particularly involving Assam and the neighbouring states. Along with local migrations and interstate migration, there is another stream of migration in the Himalaya, namely slow and continuous migration of Nepali cultivators, graziers and artisans. The exact number of such migrants is not known, but their presence influences the economic and political processes in several districts, both in western and eastern Himalaya. A detailed and systematic study of the various facets of migration in the Himalaya is called for. As part of it and of course even as a matter of general interest, a short survey will be taken of the ethnic composition of the population in the Himalaya.

SCHEDULED TRIBES AND SCHEDULED CASTES

Notionally, tribal communities are identified with reference to the stage of development in technological complexity and the scale of social and economic interaction. Intensity of interaction as well as sources of social control and management are also taken into account in the characterization of the tribal communities. The concept of the scheduled tribe is however slightly different —it is a constitutional device to enable the state to take special measures in respect of communities who have remained aloof from the main economic and political currents and cross-currents in the country, particularly during colonial rule. While in a general way the communities who notionally come within the conceptual orbit of tribals are recognized as scheduled tribes, political and administrative factors also intervene. Hence there are some tribes who are not scheduled tribes and there are many scheduled tribes who are not tribals; most of these are however marginal cases. Another important politico-cultural phenomenon is the position held by diverse hereditary groups in a system of ranking of the society in terms of ritual privileges and disabilities. This criterion has led to the identification of a large number of communities as untouchable castes or depressed classes, with a politico-legal counterpart in the recognition of communities as scheduled castes. A synoptic picture at the state level is furnished below:

State/Union Territory	Scheduled caste as % of total population	Scheduled tribe as % of total population
Jammu & Kashmir	8.26	Nil
Himachal	22.24	4.09
Sikkim	4.53	—
Arunachal	0.07	79.02
Meghalaya	0.38	80.48
Nagaland	—	88.61
India	14.6	6.94

The statement shows that there are few or no ritually disabled castes or scheduled castes in areas where the tribals are predominant. Further, it is noteworthy that in those areas, a good number of non-tribals are found, obviously many of them migrants. The position becomes clearer with an examination of the distribution of scheduled castes and scheduled tribes in the rural and urban areas.

SCHEDULED TRIBES IN RURAL AND URBAN AREAS

The statement below gives the distribution of scheduled tribes in the rural and urban areas (up to the state level):

State/Union Territory	Rural	Urban
India	96.59	3.41
Himachal	99.41	0.39
Arunachal	99.07	0.93
Nagaland	95.77	4.23
Meghalaya	91.76	8.24

The statement shows that in the country as a whole, as well as in the tribal areas, the proportion of tribals living in urban areas is a small fraction of the proportion of those living in rural areas. But a slightly different picture emerges when the proportion of scheduled tribes living in the urban areas to the total population living in those areas is examined. A synoptic picture at the state and Union Territory level is given below:

State/Union Territory	Scheduled tribes in urban area as % of total population in urban area		
	Person	Male	Female
India	1.19	1.16	1.21
Himachal	0.23	0.24	0.21
Arunachal	20.07	16.57	27.72
Nagaland	37.72	29.36	55.42
Meghalaya	45.61	40.58	51.50

The statement shows that in states like Meghalaya and Nagaland, the tribals have not been completely swamped by migrants even in the urban areas.

The proportion of urban tribal males and females to the total males and females living in urban areas brings out another interesting point: except in Himachal, everywhere (including at the all-India level) the proportion of tribal females is higher than that of tribal males to the total population of the same sex living in the urban areas. This indicates that proportionately there are more non-tribals in the urban areas in the Himalaya region without their families; they are more frequently sojourners than permanent migrants in these areas.

SCHEDULED CASTES IN THE RURAL AND URBAN AREAS

A statement giving the data is furnished below.

State/Union Territory	Scheduled caste population as percentage of total population (1971)					
	Rural			Urban		
	Person	Male	Female	Person	Male	Female
India	16.05	16.10	16.00	8.76	8.85	8.89
Jammu & Kashmir	9.30	9.00	9.54	3.68	3.59	3.78
Himachal	22.81	23.01	23.60	14.67	14.36	15.09
Arunachal	0.07	0.08	0.08	0.02	0.02	0.02
Nagaland	—	Nil	—	—	—	—
Meghalaya	0.19	0.19	0.20	1.50	1.54	1.46

When the proportion of scheduled castes to the total population in the rural and urban areas respectively is considered, it is found that the ratio between the two sets of figures is much higher than that for scheduled tribes. In fact, in one state, Meghalaya, the percentage of scheduled castes to the total population is much higher in the urban areas than in the rural. This implies that urbanization in the Himalayan region has not made the specialized services of the socially disabled caste dispensable. In fact, in Meghalaya, urbanization seems to have augmented the need for it.

Ethnic Diversities in the Himalaya

The predominance of scheduled tribes in the eastern Himalaya and the presence of scheduled castes in significant numbers in the western Himalaya, give only a glimpse of the ethnic diversity in the Himalaya. In reality the ethnic scene is highly diversified in this region. In the western sector, an extremely interesting ethnic group is the Gujar. They profess Islam; possess gotra names; their huts are scattered all over the high valleys of Kashmir. They live in these huts in summer and go down to the lower regions and even to the plains in winter with large herds of buffaloes (Bose, 1972).

In the Kargil sector of Ladakh, three distinct ethnic groups professing Islam are found. These are the Balti, Sina and Dardi, and they have a considerable number of agnates living across the border in Pakistan. In this sector, there is also a small number of Buddhists. While in the past marriage between Muslims and Buddhists occurred fairly frequently, it has now practically disappeared. It seems that such marriages were a demographic adjustment to the fluctuating needs of capital accumulation and circulation of social goods on the international borders. Ladakh proper is predominantly inhabited by Buddhists among whom polyandry prevails. In fact, all along the

Himalayan belt a number of polyandrous people are found to exist, such as the Lahulis and Kinnauras of Himachal Pradesh, Jaunsaries of Uttarakhand and Galong of Arunachal. Various explanations have been advanced by different scholars for the practice of polyandry. It seems generally to be a demographic solution for a difficult economic problem, where many households have to depend simultaneously on sheep and goat rearing, trade across the border, and upland cultivation, as sources of livelihood. In their specific ecological niche, each of these is hedged with a number of uncertainties, hence notwithstanding occupational diversities it is attempted to retain the unity of the household economy. A family based on polyandrous marriage serves such a purpose. Berremen (1977) elaborates on this line of approach. He suggests that in this region monogamy, polygamy, fraternal polyandry and fraternal polygynandry, are adaptive mechanisms by which households try to optimize the ratio of people to the resources available.

Among the other people of the Himalaya region, some of the semi-nomadic tribes deserve special mention. On the slopes of the snowy ranges of Dhaula Dhar and Pir Panjal live the Gaddis. They are Hindu and have permanent villages in the valley bottoms, where they engage in agriculture. The womenfolk remain in the villages, while their men roam with their sheep and goats in the higher meadows during summer (Bose, 1972, p. 79). The Jadhs live in the basin of Jadh Ganga or Janhavi. They mainly tend flocks of sheep and goats; but they do not live in a scattered fashion like the Gujars. They cultivate barley, wheat, *phabra* and potatoes. Another group very similar to the Jadhs who are found near Badrinath are the Bhotias of Mana. There are extensive high-level meadows around Mana where the menfolk are engaged in grazing sheep, goat, yak and cross-breeds. The Johris of Midan Valley also belong to the same category (Bose, p. 86). In the eastern Himalaya, distinct semi-nomadic pastoral groups are not found except for a few Nepali castes. On the other hand almost permanently-settled cattle rearers, belonging to various Nepali castes, have now spread all over the eastern Himalaya and the adjoining hills. They form the single majority-language speakers of Arunachal; a considerable number are also found in Nagaland and Meghalaya. In fact their rapid spread has caused political tension in some areas. Some of the indigenous people of the eastern Himalaya, like the Sherdukpens and Monpas, rear sheep and goats in large numbers, and have ritualized exchange relationships for the products of animal husbandry against those of agriculture with the people of the plains.

Apart from the communities already mentioned, the other important ones in eastern India are the Lepcha, Bhotia and Nepali castes of Darjiling and Sikkim, Bungni, Miri, Khamti, Singpho, Tangsa, Nocte and Wancho of Arunachal, the Naga group of tribes of Nagaland (including the Kanyak, Ao, Angami, Lota, Rengma), Khasi, Jaintia and Garo of Meghalaya, Kachari of North Kachar Hills and Arleng (Mikir) of the Mikir Hills. These diverse ethnic groups have distinct traditions and ways of life of their own. But at the same time a trend towards expansion of their identities by selective emphasis on particular aspects of traditions and cultures is also noticeable among them.

BIBLIOGRAPHY

BERREMAN, S. D. 1977. 'Demography, Domestic Economy and Change in the Western Himalaya'. *Eastern Anthropologist*. April–June.
BOSE, S. C. 1972. *The Geography of the Himalaya*. Delhi: National Book Trust.
ROY BURMAN, B. K. 1961. 'Demographic and Social Profile of Hills of North East India'. *Census of India 1961*.
——. 1971. 'National Movements among the Tribals'. *Secular Democracy*, Annual Number, Vol. N, Nos. 3 & 4, Delhi.
——. 1961. 'Rupa (A Village in Arunachal)'. *Census of India 1961*.
SANKALIA, H. D. 1973. 'Prehistoric and Protohistoric Period'. *Gazetteer of India* Vol. II.
SETH, S. K. 1978. 'Forests and Forestry in the Himalayan Region'. National Seminar on Resources, Development and Environment in the Himalayan Region, 10-13 April 1978, New Delhi.
SINHA, S. C. 1972. 'Tribal Solidarity Movements', in K. S. Singh (ed.) *Tribal Situation in India*. Indian Institute of Advanced Studies. Simla.
WADIA, D. N. 1965. 'Geology'. *Gazetteer of India* Vol. I.

CHAPTER 13

SHIFTING CULTIVATION AND ECONOMIC CHANGE IN THE NORTH-EASTERN HIMALAYA

I. K. BARTHAKUR

Shifting cultivation seems to have originated around 7000 B.C. through the urge of small human societies to supplement their hunting and food-gathering in the forests by the newly-discovered technique of raising food crops by planting. It then represented a new, revolutionary and efficient technology—the use of fire to clear the land of trees and undergrowth, to let in the sun, and enrich the acidic soil with alkaline ash for its cultivation. It was a world-wide phenomenon and persisted in Europe until only a few centuries ago; the burnt clearings were known in the English country-side as swiddens. It is still practised under different names in several parts of the world, particularly in the wet tropics, as it is the easiest, cheapest, and most profitable technique available to the communities in question, until their population multiplies beyond the capacity of the land to regenerate itself for the carefully planned cultivation cycle.

'Slash and burn', the colloquial description of the practice, is misleading. It creates the impression of indiscriminate forest destruction, like that ascribed to the Indo-Aryan groups who laid waste the rich forests of the Indo-Gangetic plain in their drive towards the East. The tribal people, on the contrary, were steeped in respect for nature and its harmony. Their priests and leaders carefully regulated the cycle of burning on the hill slopes. By the time the heavy rains came, the soil was covered with maize or hill millet or jungle vegetables. The cycle of *jhum*ing, as it has come to be known, in the north-eastern hills of India enabled regeneration of the forests, before the same land was cultivated again. Even as late as 1947, Charles Stonor, the Agricultural Officer of the then North East Frontier Agency India (now Arunachal Pradesh), concluded from his survey that the *jhum* cycle was as long as 25 years.

It is only within the last 30 years that the ecological balance in the north-eastern Himalaya has been shattered. The latest surveys show that in a few areas of Arunachal Pradesh—fortunately, still few and still amenable to treatment—the *jhum* cycle has shrunk to four to five years, well below the danger level. The practice is a great destroyer of forests and has an adverse effect on the ecology since it not only kills trees, but also a variety of flora and fauna, and leads to indiscriminate hunting and killing. Some animals do however manage to escape and retreat to the deeper forests. The entire character and balance of the ecosystem is changed; a totally new scheme of succession and evolution of fauna and flora emerges. Usually it is an unhappy scheme for man. 'Widespread modification or destruction of tropical forests is of concern to scientists because of their ecological diversity, complexity of structure and richness in species. Within the tropical region, intimate relationships exist between the indigenous populations and the forest, and the forest is important for soil and water conservation. Because of their extent, biomass and dynamics, tropical forests also play an important yet little understood role in the global ecological and atmospheric balance of the biosphere.'[1] It therefore becomes necessary to examine the practice of shifting cultivation in depth. Exactly what the practice consists of, why it is followed on the hills, and why it persists and what the need is to bring about a change in this practice, are cardinal questions.

Hitherto, assessments of the ecological danger to the Himalaya have concentrated almost exclusively on this traditional tribal practice of *jhum*ing. It has been condemned and new agricultural techniques have been recommended as an alternative. Yet it persists stubbornly, largely because extension workers with their new techniques make little effort to understand the origin of *jhum*ing, and the advantages it bestows on shifting cultivators or the farmer's involvement with local culture and religious beliefs.

MAJOR CONSTRAINTS—ISOLATED ECONOMIES

The hills are difficult to live in. High hills offer low returns and restrict alternatives. In the absence of road or any other transport, development of land, culture, tradition, and society takes place in utter isolation. In such areas, specialization is unprofitable, as areas segregated by high hills or non-negotiable rivers become isolated economic universes. The basic structure of these economies develops all the characteristics of desert island Robinson Crusoe[2] economies. But compared to these secluded hill economies, Robinson Crusoe had at least the advantage of higher economic standards. His wants belonged to a higher level, and he adopted better skills to achieve them. As against higher wants and knowledge, the isolated and segregated economies of the high Himalayan hills were tied to primitive standards and

[1] *Nature*, Vol. 256, p. 157.
[2] H. Speight, *The Science of Princes and Income*.
[3] *Census of India*, 1971.

knowledge: they were restricted by limited know-how, skill levels, and experience. The benefits of advancing technology and modern thought did not reach these isolated economic units and this seriously inhibited economic growth. Furthermore, contact with the adjoining and neighbouring economies which could have been of mutual advantage was also not available. Such isolation created economic constraints and established a psychological barrier which wedded the groups of people to their own customs, practices and traditions, and economic standards. They could not perceive the possibilities of broader development. All these factors influenced the general economic pattern, income, consumption, savings and capital formation, and regulated their life. The income matched what was consumed except in the case of livestock, mainly *Methon (Bos frontolis)* which they used for barter. Wood for the construction of a hut was extracted from the forests, and the labour was provided by the community. In reciprocal labour they stored labour. Of necessity many of their activities became community based—collectively they could survive. The literal meaning of the Assamese word *'jhum'* is 'collective'.

In such communities, everyone worked for self, family, and for a cohesive village, clan or tribal community. The village economy was the centre of economic and social life, each regulating the other. Thus for centuries the technology had stayed in a frozen state. Neither did patterns of consumption change over a long period. Trade and commerce were based on the scanty surpluses and meagre needs and the barter system was prevalent. The prohibitive transport costs of carrying goods on headloads along the difficult to non-negotiable paths, made trade unprofitable. Therefore, trade was restricted to essential items like salt or precious goods. The terrain coupled with absence of roads and transport facilities encouraged the formation of small and isolated villages; the density of population adjusted itself to the capacity of the difficult hilly terrain which does not sustain a high density.

In short, constraints like absence of basic economic infrastructure, lack of contact with the outside world, lack of higher technology, difficult terrain, lack of economic incentives, and above all, the absence of alternatives, kept the economies of these ancient Himalayan people, underdeveloped and poor. In the absence of more profitable alternatives there was no option but reliance on purely traditional methods of farming.

Shifting Cultivation (Jhuming)

Cultivation on the hills means cultivation on slopes. To convert slopes into flat lands requires heavy labour, know-how, and other inputs. Hill slopes when cultivated are prone to soil erosion through rain and wind. Therefore, they do not lend themselves easily to permanent cultivation of grain crops. The difficult terrain makes it extremely expensive to carry inputs for slope cultivation, and the surplus from one man's labour in the mountain cultivation is obviously small. This phenomenon also leads to low concentrations of wealth. Labour, capital and alternatives are all very scarce.

A suitable plot of hillslope is carefully selected. A number of factors like type and growth of vegetation, the texture, colour and depth of the soil, exposure of the slope to the sunshine, are taken into consideration in selection of a suitable plot. After the preliminary selection, a number of ceremonies and customs are observed to invoke divine and ancestral blessing, as well as to test the infestation of insects, pests, fungi and depredations by animals.

In many cases the services of a haruspex are also utilized for the final selection of the *jhum* site. Once the plot is finally selected, the entire village community, or in slightly modernized areas, the entire family with or without the aid of hired hands or reciprocal labour, slashes down the vegetative growth. The vegetation is cut and allowed to be dried before it is set on fire. Big trees may not be felled. The fire supplements man's efforts and clears the site for cultivation, kills fungi, insects, pests and their larvae and destroys weeds and their seeds. The ashes provide saltpetre and manure. No animal power is used by *jhumia*s for preparation of the land to be cultivated. Seeds are either dibbled in with the help of a pointed stick or broadcast. As little disturbance to the soil as possible is caused. A number of mixed crops are usually grown. Grain crops, chillies, leafy vegetables, small grains and other requirements are grown as mixed crops with a main crop predominating. The crops are continually grown, intermixed, for a number of months, and harvested whenever they mature. In this fashion most of the consumption requirements of the family are met from the swidden fields.

The method basically involves cultivating hilly tracts without terracing and without permanent investments and inputs. The natural fertility of the land is exploited. Long-rested *jhum* lands are usually fertile, and cultivation can be done on the same plot for two to three years. When the soil loses its fertility, the plots are abandoned after one cultivation and left to grow residual crops, and another slope for cultivation is opened. The cultivator may return to the same plot of land after a number of years during which period he has cultivated other plots by rotation; a number of patches of land are thus locked under a *jhum* cycle.

Disadvantages of Jhuming

The destruction of soil caused by *jhum*ing is so gradual and imperceptible to the farmer that he does not feel the need to devise methods to retain its fertility until it is too late.

Heavy rainfall washes away the exposed and tilled top soil of the hills, causing a drastic fall in land fertility and the regenerative capacity of land, upsetting the ecology of the area. The hills are denuded of forests and vegetation, while streams and brooks dry up. Once the generative top soil is washed away, it is extremely difficult and expensive to rebuild it or to restore its fertility; it is the top soil that eventually becomes impoverished in the hills as a result of generations of extensive *jhum*ing. The increase in population intensifies the pressure on land. The most important imbalance is brought about by the destruction of the equation between hunting-cum-collectional economy and *jhum*ing. As the *jhum* cycle becomes successively shorter, the greater is the speed of soil erosion.

Eventually the land which reaches the marginal fertility level begins to worry the shifting cultivator, and soon he finds himself a daily-wage labourer because fertile land is no longer available. Over time, plot by plot, hill slope by hill slope, the land gets pushed into sub-marginal strata. On such land cultivation becomes completely uneconomic and the population migrates elsewhere. When better land is no longer accessible, in the absence of alternatives, man begins to build terraces to retain the obtainable soil. Many hill slopes of the Himalaya and of many other countries, have thus been terraced.

In Nagaland, terraces are built and a number of ingenious ways devised by the people to restore fertility to the depleted soil. In many areas of Arunachal Pradesh, the felled trees and logs are kept in anticontour fashion in an effort to obstruct soil erosion. In Kameng District of Arunachal Pradesh, particularly in the Monpa area, stone walls are constructed to retain the soil. In Sikkim, Bhutan, Nepal, Nagaland, Himachal Pradesh, many other districts of western Himalaya and many countries of the world, the hill farmer has to build stone walls as high as twenty metres to hold back a few metres of soil and thus prevent the soil from being washed away completely. Erosion exposes the deep stones, which are later used for constructing terrace walls.

In olden times when land was freely available, as soon as a particular area faced population pressure, the people migrated to uninhabited and forested lands. Such freedom however, is no longer available in the north-eastern Himalayan region, where the ownership of land by individual families, tribes and villages is more or less being recognized and established.

Areas in Arunachal Pradesh where *jhum*ing has reached dangerous levels are the valleys of Eastern Kameng and the Wancho area of Tirap District. The success already attained in areas of Siang District which only a couple of decades ago had approached the same danger levels give hope for the future. Around Basar, old *jhum* fields are rapidly being regenerated. There has been considerable extension of permanent terraced and irrigated cultivation on valley slopes with the hilltops reverting to forest.

Vicious Cycle

Conceding that heavy rains help in the quick regeneration of vegetation, the fact remains that they also rapidly wash down the top soil from the hill slopes. Loss of topsoil makes the soil successively less generative, and a vicious cycle gets established, changing the ecology very rapidly. Both animals and edible plants become scarce in the forests. Yields from the swidden fields dwindle. Due to the reduced output of food, the hillman begins to lose his health and vitality. He no longer remains a chubby, sturdy, well-nourished man. He is forced to open up larger areas for cultivation in a bid to reap the basic minimum food for his growing family. Every member works harder than ever before, but obtains diminishing returns, and farmers are plunged into deep poverty.

At this stage, serious social and emotional disturbances take place, and a vacuum is apparent. Many things attract the farmers but they have no means to pay for them. This period of social and economic change is crucial for them, and it is at this stage that new types of cultivation are introduced. Sub-marginal *jhum* lands are pushed out of cultivation, and wherever possible the bench terraces which have over time gained in fertility and are more productive than the marginal *jhum* lands, are drawn into cultivation; sedentary cultivation is finally prac-tised. At this juncture, shifting cultivators need understanding and careful, trained handling to conduct them across these confusing crossroads.

Farmers in hilly regions all over the world have passed through the *jhum*-ing stage. In certain areas the hills were completely denuded, whereas, in other places the denudation process was halted midway and terraces were built or alternatives were found and the ecology of the area protected. Horti-culture trees replaced the natural forests. In the north-eastern Himalayan ranges however, very serious damage has not yet occurred, compared to the western Himalaya. But if the natural course of events is allowed unaltered, the obvious results will ensue. To protect the ecology and to provide for the growing needs of a growing family of man, a balance between the rights of nature and the needs of man has to be struck. Therefore, it is imperative that the reasons and the circumstances which encourage shifting cultivation be understood in depth.

Advantages of Jhuming

The disadvantages mentioned are, however, not apparent to the man involved. He is not aware of them and even if he were told that after a hundred years his land would be impoverished, he might not change. What else can he do 'Now' to live? To him 'Now' is more important than a hundred years hence. He continues to reap the advantages from the swidden type of cultivation, because that is the only way open to him. He has not seen other denuded hills and impoverished lands, and may not comprehend that the *jhum*ing done by people of the distant past had created the impoverished hill slopes of many countries of the world.

Jhum fires quickly render dense forests and foliage fit for growing crops. Fire is thus a great labour-saving device. The ashes correct the soil acidity, and the admixture of ashes with the soil makes the soil more fertile. The soils of hill slopes of high rain zones are generally acidic, and are thus partly neutralized by the alkali content of the ashes. Further, fire clears the area of extensive preponderance of fungi, insects and pests. Not only are the insects and pests of the *jhum* fields destroyed along with their larvae and eggs, but the *jhum* fires which are usually lit at night and continue for a few days, attract insects and pests from the surrounding forests and destroy them too. This rids the surrounding area and the fields of these pests which would have destroyed the crops. The clearance of insects and pests from the encompassing area of the *jhum* plot creates a protective belt and keeps the crops safe from the heavy attacks of insects and pests. In many parts of the country, it has been seen that stoppage of *jhum* fires has led to an enormous increase in the incidence of insects and pests. In some areas, the attacks of insects and pests have been so heavy that almost 50 per cent of the crops were lost. Moreover, it is believed that burning also retards the growth of weeds in the *jhum* clearing by destroying the roots of the tubers and seeds of the weeds, and rendering them dormant for some time. This reduces the problem of weeding to an extent.

Experience shows that the seeds start germinating when the cold hill soil is warmed by the sun at the end of April or May. But when a particular patch of hill slope, which has been cleared for *jhum* and has sustained the large *jhum* fires, is sown after the ashes have just cooled, the seeds in the fire-warmed soil sprout within a week or so, that is, in the early part of April or even towards the end of March. This affords several clear advantages to hill cultivation. The crop has a longer maturing period and therefore, a greater yield. The seedlings derive the benefit of pre-monsoon showers and become well rooted in the heavy top-soil of the clearing. Otherwise, the monsoon downpour washes away the ungerminated seeds along with the top soil, leading to an extremely thin stand of crops.

The shifting cultivator thus reaps a number of very tangible technological advantages like lower input of labour, elimination of pests, insects and fungi, neutralization and fertilization of soil, less time and labour spent on weeding and early cultivation of crops. Without firing, he would have laboured harder and reaped less. No wonder that to the *jhum* cultivators the grains harvested from *jhum* fields are the tastiest.

Also, the hill man finds it extremely uncomfortable to work on flat land where he has to bend double. Perhaps due to this single factor, almost all the tribes who practise shifting cultivation prefer to cultivate hill slopes rather than valleys. It is only when the hills become unproductive that the valleys are brought under cultivation, but before this has happened, when in need, they tend to sell away the valley lands at nominal prices and realize their mistake rather belatedly.

Furthermore, an advantage of the *jhum* system which represents a compromise between pure hunting and food gathering on the part of the man and the very limited kitchen-garden cultivation practical for the woman, is that during the very process of movement to, and working of the *jhum* fields, the trapping of animals, even smaller ones like squirrels, or birds, can be carried on. The jungle inside the *jhum* clearance, both before and after its burning, and the forest immediately surrounding the *jhum* fields, yield a variety of vegetables, mushrooms, and tubers, not to mention the firewood which can be conveniently collected by children, to provide a more varied diet through diversified activity, contributing towards a happier and more balanced mental make-up than single-minded concentration on cultivating permanent terraced fields.

Shifting cultivators usually grow mixed crops on *jhum* fields. They harvest continuously for a long period and obtain an admixture of crops one after the other which provides them with almost everything they need for daily use. Whatever extra is required is collected from the forests. They sell or purchase very little. Moreover on *jhum*s less labour input per day is required over an extended period of time, compared to the intensive labour input required in the case of settled cultivation, particularly of irrigated terraced paddy fields. According to a study[4] conducted in 1972 on the input–output of plots under *jhum* cultivation vis-à-vis permanent bench terraces, in the hills of Arunachal Pradesh, the soils have not yet reached the eroded state. Similarly, the total yield from the *jhum* plots was found to be higher than the new permanent terraced fields cultivated without extraneous inputs.[5]

New *jhum*s on virgin or long-rested lands are highly productive. In many cases when terraces are built, the top soil is cut away or the generative soil is buried too deep under the 'half-cut half-filled' terracing method, thereby making the terraces less fertile or even unfit for cultivation. Such terraces do not acquire the required fertility for a long time, but over time and with inputs, do improve in fertility, while the reverse is the case with *jhum* lands.

In summary, it may be said that with reduction in the *jhum* cycle, soil fertility quickly decreases, resulting in further reduction in the cycle. Leaching, erosion and loss of fertility takes place rapidly, and the yield per unit of land becomes progressively lower. All this further aggravates the situation, and makes the shifting cultivator increasingly poor in spite of his putting in much greater labour. He is slowly forced to sell off or eat up his livestock. He and his family begin to swarm in the unskilled daily-wage market. His health level falls, and he learns the hard lessons of poverty and deprivation. He succumbs easily to disease. Once the stage is reached where the *jhum*s

---

[4]I.K. Barthakur,'The Problems of Agricultural Development in Arunachal Pradesh (with special reference to Subansiri District)', Ph.D. thesis, unpublished.

[5]cf. Village survey, Agro-Economic Research Centre, Jorhat, Assam, India. Similar results were found there too.

are no longer productive, events follow so swiftly, that the cultivators are
taken unaware. Not understanding the reasons fully, as a defence they
begin to refuse to try anything 'new'. To their simple minds, it must be the
'new' which is causing pain. In the olden days they had plenty and their
life was a happy song.

At this stage shifting cultivators need understanding and careful handling
to make them understand the forces of change. Trade and commerce can
give added value to the otherwise free gifts of nature, and instil new life into
stagnant economies. The population may tend to form bigger villages, and
there is greater diversification of occupation. Migration, and such diversification can lengthen the rest period of *jhum* plots.

Wherever new roads are built, the forests immediately acquire economic
value, and the loggers promptly begin their activity. However, the overall trend
becomes favourable to permanent cultivation. Waste lands begin to be cultivated and hills are terraced. Agricultural and other permanent investments
are introduced. The cropping pattern undergoes a change and orchards and
other cash crops appear on the slopes. And in the areas beyond the direct
influence of roads, population pressures tend to lighten due to the emigration of population to the road points. There the forests begin to regenerate.
The growing villages, which had earlier due to their small size remained
unqualified for development aids like schools, medical aid, water supply,
and electricity, now receive these benefits, which in turn attract more settlers.
The very familiar process of development of townships begins to unfold.
The place no longer remains primitive.

This stage, of systematically developing the hills in a planned manner,
and constructing correct terraces with the use of inputs to enhance fertility
of the terraces is important. Even the partial shifting of pressures of food
requirements from the *jhum* lands to terraces and imports, would help in
safeguarding the ecology.

Cutting correct terraces is vital. Otherwise, the entire terrace-cutting programme may boomerang and cause untold hardship to the progressive
jhum cultivator. Wrong terrace-cutting either buries the top generative soil
too deep for the shallow roots of the grain crops to reach, or throws away
the top soil. In both situations the farmer would not receive any return
for his initiative, enthusiasm, labour, seeds and other meagre investments.
He and his family may face the risk of starvation. The neighbouring farmers
learning of his sad experience may shy away from the programme of terrace
cultivation.

There are many instances where wrongly-cut terraces built with the aid
of Government subsidy and persuasion were subsequently left uncultivated
by people for a number of years. This is because badly-cut terraces, especially
those constructed on high slopes, displace fertile top soil and yield very low
returns. This encourages the cultivator to continue shifting cultivation
because the productivity of traditional *jhum* fields is relatively higher than

that of newly-cut terraces without appropriate inputs in the initial years. The terraces yield less return because the advantages of burning the vegetative growth and application of ashes as fertilizer are not available in terraced fields. In a study[6] conducted in a district of Arunachal Pradesh hills, the villagers complained that even after leaving a freshly constructed terrace for the period equivalent to a local *jhum* cycle, the comparative productivity remained low.

Water availability and water management for irrigating the terraces, and preparing the soil fertility level for reaping more than one crop are fundamental to successful terracing on eroding hill slopes. These are usually not available. Many of the mixed crops which the *jhum* cultivator grew in his *jhum* fields cannot be grown in wet terraces. A number of new problems such as the menace of insects, pests and fungi emerge because these are no longer destroyed by the annual *jhum* fires. The building of terraces and cultivation of terraces is comparatively labour-intensive and an expensive process.

Therefore, before an attempt is made to encourage shifting cultivators to adopt terrace cultivation, it is imperative to fully educate and involve them in the reasons and correct method of doing so. Cutting the lowest terrace first, and building an appropriate bund capable of holding a colossal amount of soil (which seems to swell when the hill slopes are cut for terraces), is likely to help in retaining the top soil. The top soil of the second terrace can easily be thrown on to the first terrace. Similarly, the top soil of the third terrace is received by the second terrace and so on. However, the top-most terrace in this process is ultimately bereft of its top soil.

The Indian Council of Agricultural Research have recently conducted research into *jhum*ing from their newly-established (1975) complex for the North-Eastern Hill Region. In their paper, 'Shifting Cultivation in the N.E. Region', the following reasons are suggested for the failure to attract *jhum* cultivators into permanent cultivation:·

1. The new settlement cuts into their traditional cultural life abruptly.
2. They are not used to cultivation on terraces, and the use of bullocks and implements.
3. They find production low on the terraces in the first year, due to removal of top soil while developing terraces.
4. The production technology for terraces, water management, water conservation practices, etc., are also not properly developed for the region.

This assessment however ignores the most important factor—the cultivator's own perception of reality, and the lesson taught by communities like the Angamis, Aos, Tangkhuls and Apatanis, who long ago developed their

[6]Barthakur, 'Problems of Agricultural Development in Arunachal Pradesh'.

own expertise in permanent terrace cultivation, but yet retain an element of forest for *jhum*ing, hunting, trapping, and fuel-gathering to provide the essential supplement to their diet and community needs. All this indicates the need for a practical compromise between *jhum*ing and terraced agriculture on hill slopes, and recognition of the genuine difficulties for the communities concerned.

Indigenous Methods of Enriching Terraces

In the absence of road and communication facilities in the far-flung Himalayan hills of the north-eastern region, where inputs are difficult to introduce, the indigenous methods of enriching the terraces followed by some tribes appear attractive. In the Angami, Ao and Thangkhul areas of Nagaland and Manipur states, livestock is tethered on the newly-cut terrace and livestock droppings are used to improve the soil texture and fertility. Terraces as a rule are submerged in water for as long a period as possible before sowing and also after the harvest. This continuously enriches them with deposits of the top soil carried from elsewhere and simultaneously assists the process of decomposition that brings about a beneficial change in soil composition. The submersion of the terraces for the greater part of the year seems to help build a particular level of micro-organism that brings normally insoluble plant nutrients into solution.[7] Further, at the time of the construction of the terraces on a high slope, the Naga tribes often keep the original top soil in heaps and spread it back on the terrace after levelling is completed. The Monpas of the western Kameng district of Arunachal Pradesh decompose oak leaves in compost stacks with night soil and rubbish, which they spread later.

It is significant that some more advanced tribes of the Naga areas, who have adopted extensive permanently irrigated cultivation for centuries, have yet not abandoned *jhum*ing totally. They have sought to improve upon their permanent cultivation by burning the undergrowth by a method which resembles, but which is not totally identifiable, with the process of *jhum*ing.

In the Angami Naga area, where construction of permanent irrigated fields is not immediately possible, certain hill slopes are gradually developed into permanent rain-fed fields by the device of contour bunding, first with the logs yielded by the *jhum* and sometimes by supplementary rough dry-stone masonry where stones become available on the hill slopes. In such transitional fields certain trees, notably the oak and alder, are left standing, as the fallen leaves act as compost. The yield of these leaves is increased deliberately for this purpose by cutting the upper portions of the trees for firewood, and by pruning the new shoots annually under the process known as pollarding.

[7] *Hunger Signs in Crops*, American Society of Agronomy, p. 11.

The Approach to Change

In making a shift from *jhum* to permanent cultivation, the whole problem
of land tenure and the rights of transfer of land arise, leading to something
almost like a social revolution in the village. The very word *jhum* means
'collective'; an area which might be fit for permanent irrigated terracing
may well have originally constituted an area whose cultivation was sancti-
fied by tradition by a particular clan, community or family in rotation over
the years. If properly developed for permanent terraces, the trend towards
individual ownership of irrigated terraces naturally becomes irresistible, and
a scramble may arise to buy such terraces even outside the traditional orbit
of ownership. The new system of irrigated terraces may bring in its wake the
problems of conflict between operational and traditional ownership. A
special responsibility will devolve on the agents of change for preparing the
community to face, overcome, or sometimes to live with the stresses and
strains that such a transition produces. The abandoning of *jhum*ing will
change the present way of life; what the new values will be has to be ascertain-
ed, and sociological changes must be ushered in alongside. This is indeed a vast
responsibility. The central theme remains. Any other type of agriculture has
to be economically more profitable than *jhum*ing and socially more beneficial, to make the people change over from *jhum*ing. It must not only be
theoretically more profitable than *jhum*ing but seen to be more profitable in
the fields by the villagers. To be successful, any programme of change must
be viewed through the eyes of the person who is to undergo the change. Let
us look at the practice from his point of view before suggesting a line of
action to restore the ecological balance without at the same time destroying
his own; for the human population of the environment surely has an equal
right to consideration in the totality of the environment itself.

There is increasing awareness, education, and knowledge of the wider world
among the communities involved. Their perceptions are already tending to
confront the orthodox planners, with a world of 'different realities'. They
see the prime cause of devastation of the ecology to be the organized denuding
of forests by 'intruders' from the plains, and of siltage of rivers on account
of the blasting for roads and construction sites. A reconnaissance largely
substantiates their perception that the main damage is along the roadside
belts and new townships and less in the interior *jhum* lands. In their perception, even the rapid reduction of the *jhum* cycle is partly at least attributable
to the shrinkage of their traditional land caused by forest reservations and
restrictions. Their perception demands satisfaction that alternative techniques such as permanent terraces actually produce at least an equal food yield
to their traditional *jhum*s; that their nutrition does not suffer from denial of
the produce which they used to collect from the forests; that cash crop programmes actually bring in cash and a better standard of living for themselves
and not merely to middlemen or corporations who in the name of 'cost

efficiency' reduce them to daily wage labourers, and above all, that their quality of life is enriched and not impoverished by change. At Darchai in Tripura the people themselves, a generation ago, evolved a system of planting abandoned *jhum*s cyclically with citrous trees. Similarly, P.D. Stracey, an experienced and innovative forest officer, had evolved a scheme for replanting abandoned *jhum*s with quick-growing high-yielding vegetation to provide cover against erosion and increase fertility for the next cycle, pending an increase in permanent irrigated terracing. Hill villagers are traditionally well aware of the need to conserve forest cover near water sources and take well to horti-cultural programmes, *provided* such programmes have a built-in marketing or processing component. Flat valley lands or gentle irrigable slopes have to be carefully and systematically surveyed and identified and irrigation facilities coordinated with terrace cutting. The loss of crop while cutting terraces has to be compensated through food for work programmes. Grazing facilities for village herds at different altitudes in different seasons and their passage through the area needs to be carefully provided for. This requires hard thinking and planning in detail down to village levels, if the hill people are not to be impoverished by development programmes oriented purely to-wards commercial forestry. Above all, it is the day-to-day decision maker, extension worker, the man on the spot, who is the pivotal agent for change.

Clearly, all this argues the need for considerable rethinking at policy planning levels, and an abandonment of departmental approaches is required. If in the modern world war is too serious a matter to be left to soldiers, ecology is too important to be left to forest departments. The world trend is towards the evolution of integrated land-use policies, described by Kenneth King of the FAO as agro-silviculture or agro-pastoral-silviculture. Traditional *jhum*ing is after all only cyclical agro-silviculture and the modern approach to it should derive sustenance from the experience of communities who are evolving beyond it, within a multi-package integrated programme of rural development in the hills. Policy planning therefore has to be highly sensitive to the new needs, provide correct leadership, thought, knowledge and involve the people in bringing about a harmonious development of Himalayan man and his mountainous environment.

BIBLIOGRAPHY

ELWIN, V. 1959. *A Philosophy for NEFA*. Second edition. Shillong: North East Frontier Agency.

BARTHAKUR, J.K. 1967. 'A Note on the Backwardness of Assam Hills'. *Common Pers-pectives of North East India*. Calcutta: Pannalal Das Gupta.

———. 1978. 'Demography and Socio-Economic Processes in Arunachal Pradesh'. Ph.D. thesis, Gauhati University.

BISWAS, J. 1966. *JARA (A Dafla Village Monograph)*. New Delhi: Publications Division, Government of India.

BISWAS, S. and GHOSH, A.R. 1977. 'Impact of Shifting Cultivation on Wild Life of Meghalaya'. Shillong.

BOWER, U. G. 1953. 'The Hidden Land'. *Stagnation to Growth*. London: John Murray.

DANTWALA, M. L. 1970. 'Presidential Address to the Indian Agricultural Society at Gauhati'. Bombay.

Dhebar Commission. 1961. *Report on the Scheduled Areas and Scheduled Tribes Commission*. New Delhi: Publications Division, Government of India.

WILSON, G. E. E. 1954. *The Social Economics of Agriculture*. Third edition. New York: Macmillan.

GOSWAMI, P. C. 1971. 'Problems of Agricultural Development in Tribal Areas'. *Indian Journal of Agricultural Economics*, Vol. XXV No. 3. Bombay: Indian Society of Agricultural Economics.

HAIMENDORF, C. V. 1962. *Apatani & Their Neighbours*. London: Routledge and Kegan Paul.

Indian Council of Agricultural Research. 1964. *Agriculture in Ancient India*. New Delhi: Publications Division, Government of India.

KATJU, K. N. 1953. *Rural Development Through Self-Help*. New Delhi.

LEACH, E. R. 1954. *Political System of Highland Burma*. Cambridge, Mass.

MOSHER, A. T. 1969. *Creating a Progressive Rural Structure*. New York: The Agricultural Development Council.

PANT, S. D. 1935. *The Socio-Economics of Himalayas*. London.

PIERCE PHILLIPS, JOHN. 1961. *The Development of Agriculture and Forestry in the Tropics—Patterns, Problems and Promise*. London.

SCHLIPPE, PIERRE DE. 1956. *Shifting Cultivation in Africa: The Zande System*. London: Routledge & Kegan Paul.

SHAKESPEAR, L. W. 1914. *History of Upper Assam, Upper Burma and North Eastern Frontier*. London.

STRACY, P. D. *Bringing the North East Frontier Agency Alive*.

STYENSON, H. N. C. 1943. *The Economics of Central Chin Tribes*. Bombay.

SPEIGHT, H. *The Science of Prices and Income*.